Student Solutions Manual
for

Moore, McCabe, Alwan, Craig, Duckworth's
The Practice of Statistics
for Business and Economics

Third Edition

Christa Sorola
Purdue University

W. H. Freeman and Company
New York

Chapter 1: Examining Distributions

1.1 Answers will vary.

1.3 Exam1 = 71; Exam2 = 80; Final = 79.

1.5 Each case in the data set represents an apartment available to rent. There are five variables: monthly rent (quantitative), free cable (categorical as answers would be "yes" or "no"), pets allowed (categorical), number of bedrooms (quantitative), and distance to campus (quantitative).

1.7 **a)** See bar graph below. The bars are displayed from largest percentage to smallest percentage to highlight the company that has the greatest market share.

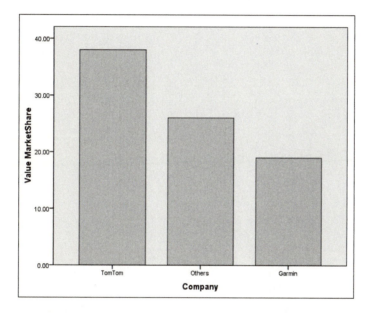

b) Garmin has its headquarters in the United States, so one would expect this company to have a greater percentage of the U.S. market share. TomTom is headquartered in Europe, so you would expect their market share to be higher in the European market.

1.9 Since the width of the intervals for the histogram need to be equal in size, you would replace the < sign with a ≤ sign and the ≤ sign with a < sign. If any students earned a score of 60 or lower, a new interval should be added for scores in this range: 50 < rate ≤ 60.

1.11 Stemplots will vary but should have the characteristics described.

1.13 **a)** Each case represents an employee of the company. **b)** Labels include employee identification number, last name, first name, and middle initial; categorical variables include department and education; quantitative variables include number of years with the company, salary, and age. **c)** Spreadsheets will vary. Column titles should correspond with variable and label names from part (b). Each row should represent a different employee.

1.15   **a)** is quantitative, as age is a numerical value. **b)** is categorical, as "yes" or "no" will put the student into one of these categories. **c)** is categorical, as there are three categories available for the responses. **d)** is quantitative, as this will be a dollar amount. **e)** is quantitative, as height is a numerical value **f)** is categorical, as either a "yes" or a "no" is expected.

1.17   Answers will vary.

1.19   Orange is the least popular color, followed by brown. Approximately the same number of people selected purple, yellow, and gray as their least favorite colors, and this percent was much lower than the percent that selected either orange of brown. No one selected either black or blue as their least favorite color.

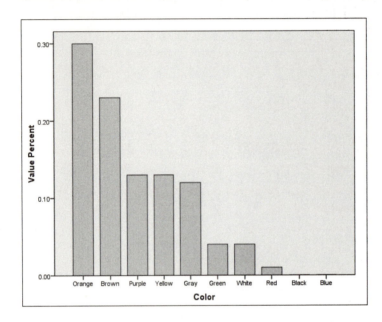

1.21   See bar graph below. Alternate orderings are acceptable with appropriate reasoning. This order was selected to highlight the specific type of phone that was replaced from most to least common. The other categories were then noted at the end of the bar graph.

1.23   **a)** See bar graph below. **b)** The bar graph clearly shows that Google Global has a huge portion of the market share compared to the other search engines.

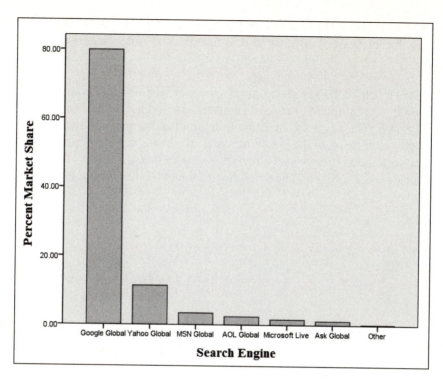

1.25 **a)** See bar graph below. **b)** The graph below shows that the United Kingdom and Canada have the greatest number of Facebook users with more than twice the number of any other country. Turkey, Australia, Colombia, Chile, and France make up a second tier of countries with the remaining countries showing less than half the number of users as any of these five. Italy had the smallest number of users.

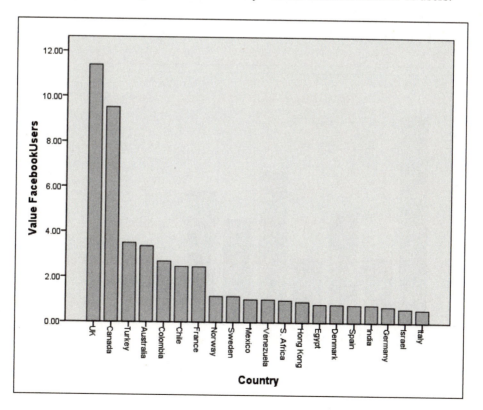

d) With data values as spread out as these, a stemplot is not very effective. A histogram would allow for intervals of different widths than the 10% intervals created by a stemplot.

1.27   **a)** See histogram below. **b)** See stemplot below. **c)** Both graphs show an area of higher concentration in rates in the 4% range. The histogram highlights that this range is in the 4.5% to 5% range. The advantage to using a stemplot is that you can see the actual data values as well as the shape of the graph; however, the histogram highlights the area of maximum concentration more vividly than the stemplot does. Answers from students may vary but should indicate an understanding of the two types of graphs

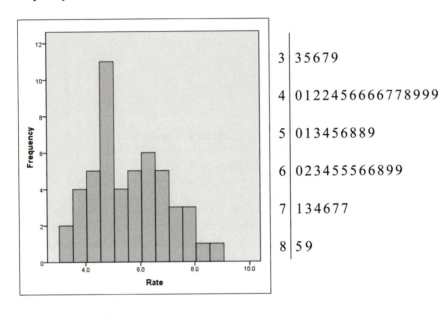

```
3 | 35679
4 | 0122456666778999
5 | 013456889
6 | 023455566899
7 | 134677
8 | 59
```

1.29   **a)** Luxury car percents.

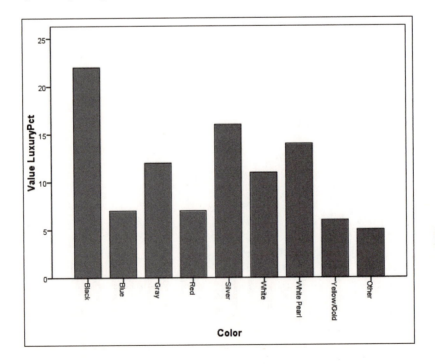

b) Intermediate car percents.

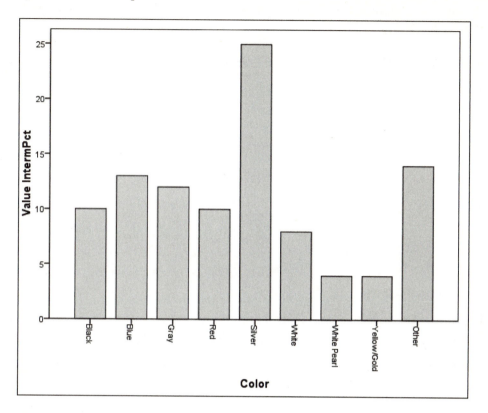

c) Combined bar graph (answers will vary).

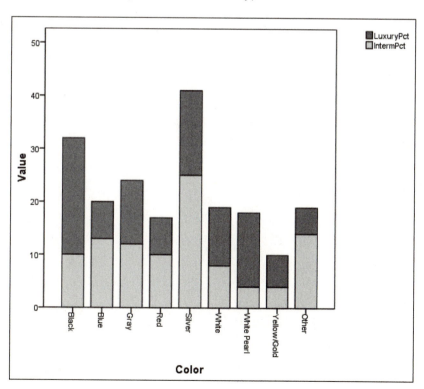

1.31   **a)** Alaska has 7.0% and Florida has 17.0% older residents. **b)** The distribution is unimodal and symmetric with a peak around 13%. Without Alaska and Florida, the range is 8.8% to 15.5%.

1.33   **a)** The stemplot shows a unimodal, strongly left-skewed distribution.

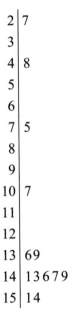

```
 2 | 7
 3 |
 4 | 8
 5 |
 6 |
 7 | 5
 8 |
 9 |
10 | 7
11 |
12 |
13 | 69
14 | 13679
15 | 14
```

b) A histogram could also be used to show quantitative data; however, the stemplot has the advantage of showing the actual data values in addition to the shape of the graph. A histogram might also show fewer intervals than the stemplot for this small of a data set; however, this might cause a loss of information on the shape of the graph at the higher percentages.

1.35   The histogram below shows a unimodal distribution that is skewed to the right. The area of greatest concentration is from 33.3 barrels to 50 barrels. The boxplot shows that this distribution also has four high outliers at 118.2, 156.5, 196.0, and 204.9 barrels.

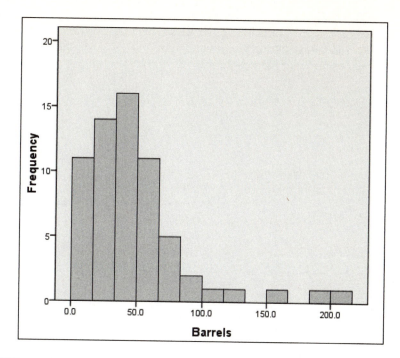

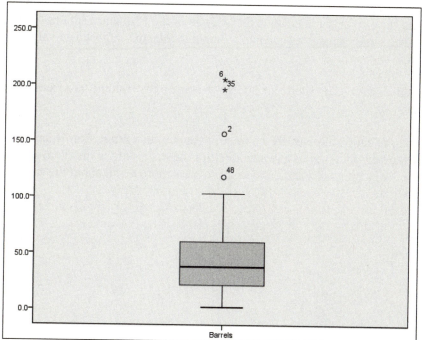

1.37 The count of service calls is not an appropriate measure to report when comparing reliability between Brand A and Brand B dishwashers. The total number of owners of Brand A and Brand B included in the study are very different (13,376 vs. 2942). **b)** A better measure of reliability might be the percentage of owners of each brand that requested a service call. Comparing these percentages shows that 22% (2942/13,376) of Brand A owners requested a service call, while 40% of Brand B owners requested a service call. It appears that Brand B has a much lower reliability than Brand A.

1.39

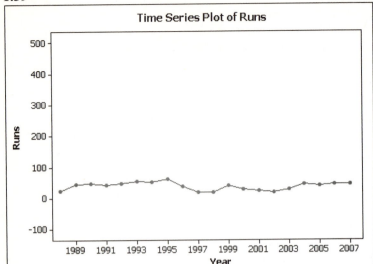

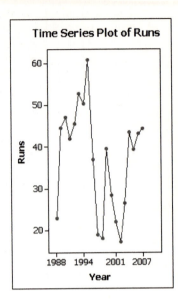

1.41   The mean service time is 196.575 seconds.

1.43   When the data values are ordered from smallest to largest, the middle value is 40 days. Without the outlier, the median is the average of the 12th and 13th observations: 36.5 days. The outlier causes the median to move to the next largest data value, an increase of 3.5 days.

1.45   The median first-exam score is 82.5, the average of the middle two exam scores of 80 and 85.

1.47   The boxplot shows a relatively symmetric graph with a range from 4 days to 77 days. While the boxplot gives us a general feel for the nature of the distribution, in this case the stemplot is much more instructive as it provides more detail about the shape of distribution.

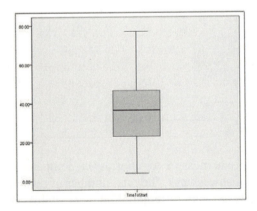

1.49   The standard deviation without Suriname is 18.6. The standard deviation with Suriname is 132.6.

1.51   **a)** The mean and standard deviation are 196.575 and 342.022. **b)** The five-number summary is minimum = 1, $Q_1 = 54.5$, median = 103.5, $Q_3 = 200$, maximum = 2631. **c)**

The distribution has quite a few outliers and is skewed to the right; therefore, the five-number summary does the better job of describing the distribution.

1.53 **a)** The median is 5.5%, Q_1 = 4.0%, and Q_3 = 6.9%. **c)** With the outliers removed, the median is 5.5%, Q_1 = 4.0%, and Q_3 = 6.5%. The median and the quartiles changed only slightly with the removal of the outliers.

1.55 **a)** The histogram and boxplot are shown below. The histogram appears to be unimodal and skewed to the right with one very low outlier; however, the boxplot shows that there are quite a few outliers. The outliers would make it hard to determine the shape of the non-outlier data using a histogram.

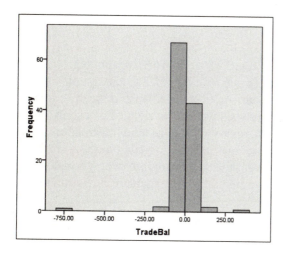
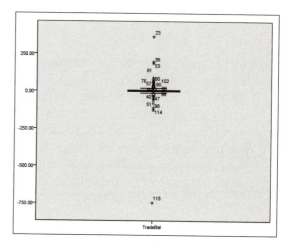

b) Countries with trade balances below –17.20 or above 21.60 would be considered outliers. The low outliers include the United States, Spain, United Kingdom, Italy, Australia, Greece, Turkey, France, Romania, South Africa, India, Portugal, and Poland. The high outliers include Taiwan, Malaysia, Canada, Sweden, Algeria, United Arab Emirates, Singapore, Kuwait, Norway, Netherlands, Switzerland, Russia, Saudi Arabia, Germany, Japan, and China. **c)** See the histogram below with the outliers removed. Now it is easier to see the shape of the interior portion of the data set. These data are also unimodal and skewed to the right, but more intervals are shown with substantial amounts of data than seen previously. **d)** The distribution is unimodal and skewed right with 29 outliers. The median is –$0.5 billion with a range from –$747.1 billion in the United States to $363.3 billion in China. When the outliers are removed, the median of the remaining data points is –$0.6 billion with a range from –$12.6 billion to $19.9 billion. In either case, it is clear that more than half of the countries shown import more than they export.

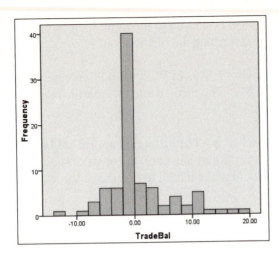

1.57   **a)** The mean is 6.93%, and the standard deviation is 3.18%. **b)** The five-number summary is minimum = 3.5%, Q_1 = 4.2%, median = 6.8%, Q_3 = 8.0%, maximum = 13.6%. **c)** See boxplot below. **d)** Both the comparison of the median and mean as well as the boxplot below show that this data is strongly skewed to the right. To describe the center and spread of this distribution, it would be appropriate to use the information from the five-number summary. Since there are only a small number of data points, and since they are rather spread out, we would probably want to use a histogram over a boxplot to show the graph of the data. In addition, there is potential for the standard deviation to be larger with smaller data sets.

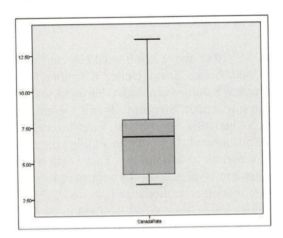

1.59   **a)** The mean is 48.25 barrels, and the standard deviation is 40.24 barrels. **b)** The five-number summary is 2.0, 21.5, 37.8, 60.1, 204.9. **c)** See boxplot below. **d)** The boxplot shows that this distribution is skewed to the right with outliers. The five-number summary is the best way to describe the center and spread of the distribution, as they are not as greatly affected by the outliers or skewness as the mean and standard deviation are.

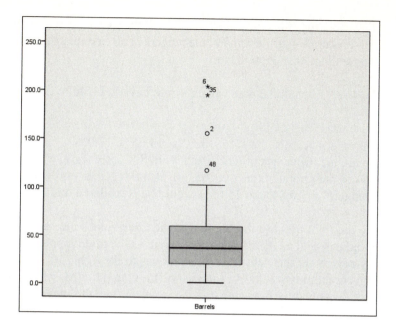

1.61   **a)**

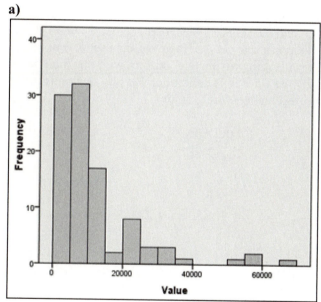

b) The mean for this distribution is $12,143.95, and the standard deviation is $12,421.48. The five-number summary is $3338, $4589, $7558.5, $13,416, $66,667. This distribution has a number of high outliers, so the five-number summary would be the appropriate numerical summary. **c)** Answers should mention that the graph is strongly skewed to the right and that the distribution contains nine high outliers. Answers should also mention the median of the distribution as the measure of the center. Other information may be included as well.

1.63   **a)** The mean of this distribution is 4.76%, and the median is 4.70%. Without the outlier, the mean is 4.81%, and the median is still 4.70%. When the outlier is excluded, the mean increases, but the median remains unchanged. **b)** The standard deviation is 0.752 but decreases to 0.586 when the outlier is removed. The first quartile moves from 4.3 to 4.35 when the outlier is excluded. The third quartile remains at 5.0 with or without the outlier.

c) Removing the outlier from the data has an impact on both the mean and standard deviation. The median and third quartile remain unchanged. The first quartile shifts up one position and so has a slight change in value.

1.65   Answers will vary; however, the middle data values would have to be quite far apart for this to happen.

1.67   **a)** The five-number summary is 7.0%, 12.15%, 13.10%, 13.55%, 17.0%.
b) Florida and Alaska are identified as outliers. Utah, with 8.8%, Georgia, with 9.9%, and Texas, with 10.0% of their populations over 65, are also flagged as outliers since any percents outside the range from 10.05% to 15.65% would be considered outliers.

1.69   **a)** The mean moves to the right because the average of the three values increases as the value of one of them increases. The median does not move as the middle point is still in the middle, regardless of how far the right-most point moves to the right. **b)** The mean moves to the left as this point moves left. As soon as the moving point passes the right-most stationary point, the median becomes the moving point. When the moving point passes the left-most stationary point, the median becomes this left-most stationary point and remains there regardless of how far the moving point moves to the left.

1.71   The mean for data set A is 7.5, and the standard deviation is 2.03. Data set B has the same mean and standard deviation as data set A. However, data set A is unimodal and skewed to the left, while data set B is approximately uniform with one high point that is just short of being an outlier by the 1.5 × IQR definition. See the stemplots below. Data points were rounded to the nearest tenth before graphing.

DATA A		DATA B	
3	1	5	3 6 8
4	7	6	6 9
5		7	0 7 9
6	1	8	5 8
7	2	9	
8	1 1 7 8	10	
9	1 1 3	11	
		12	5

```
 1 | 0 0 0 5
 2 | 7
 3 | 0 0 2 2 7 8 9
 4 | 3 6 6 6 7 9
 5 | 0 0 1 3 4 5 7 7 8 8 8
 6 | 0 1 5 6 8 9
 7 | 0 2 2 3 5 8 9
 8 | 0 0 3
 9 | 4
10 | 4
11 | 6
12 | 8
13 | 8
```

1.73   **a)** The mean salary is $79,375. All the employees (not counting the owner) earn less than the mean. The median salary is $30,000. **b)** The mean increases to $91,875. This increase in the owner's salary does not affect the median.

1.75   **a)** Allow all four numbers to be the same. **b)** The numbers 10, 10, 20, and 20 give a standard deviation of 5.77. **c)** There are multiple choices for answer (a). The standard deviation measures the variation about the mean. Choosing numbers all the same gives a standard deviation of zero. There is, however, only one choice for answer (b). Choosing an equal set of numbers as far from the mean as possible will give the largest standard deviation.

1.77   The trimmed mean is $10,262.12, while the mean of the entire data set is $12,143.95.

1.79   Sketches will vary but check to see that students understand the difference between *symmetric* and *skewed*.

1.81   **a)** The mean is at point C and the median is at point B. The mean is greater than the median because the distribution is skewed right. **b)** The mean and median are equal, and both occur at point A because the distribution is symmetric. **c)** The mean is at point A and the median is at point B. The mean is less than the median because the distribution is skewed left.

1.83   **a)** 2.5%. **b)** Between 64 and 74 inches. **c)** 16%.

1.85   Since 99.7% of the scores are within three standard deviations of the mean, a range for the central 99.7% of these scores would be (419, 725).

1.87   The proportion less than 600 is 0.7088. The proportion to the right of 600 is 0.2912.

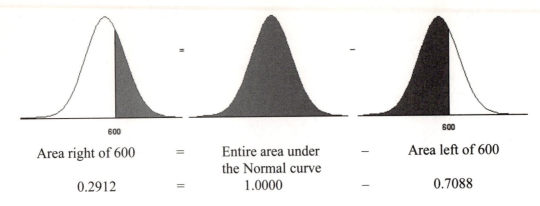

Area right of 600	=	Entire area under the Normal curve	−	Area left of 600
0.2912	=	1.0000	−	0.7088

1.89   A score of 656 would be required to be in the top 5%.

1.91   **a)** Instead of following a straight line, as the Normal quantile plot would do in the case of a Normal distribution, this plot shows a curve that increases gradually at the beginning and quite steeply at the end. **b)** In Figure 1.28, the upper portion of the curve follows a straight line. In Figure 1.29, the upper portion of the curve flattens out after a steeper increase. For right-skewed data, the Normal quantile plot will show the highest points below the diagonal. For left skewed data, the Normal quantile plot will show the highest points above the diagonal. For symmetric data, the points will follow the diagonal fairly closely, even at the ends.

1.93   **a)** and **b)** See graph below. **c)** Changing the mean shifts the curve along the *x*-axis.

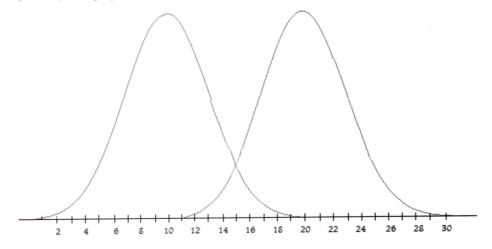

1.95   Answers will vary. See examples below.

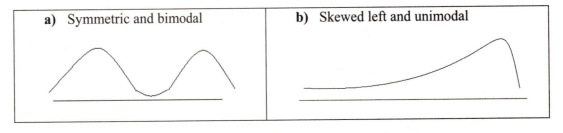

1.97   **a)** According to the 68-95-99.7 rule, 68% of the data should be in the interval from 5232 to 23,362 words, 95% of the data should be in the interval from –3833 to 32,427 words, and 99.7% of the data should be in the interval from –12,898 to 41,492 words. **b)** It would not be reasonable for a woman to speak a negative number of words, so these intervals would not make sense for this situation. In addition, if a woman spoke 41,492 words in a day, she would be speaking an average of one word every 1.39 seconds for 16 hours of the day (assuming 8 hours of sleep). **c)** According to the 68-95-99.7 rule, 68% of the data should be in the interval from 5004 to 23,116 words, 95% of the data should be in the interval from –4052 to 32,172 words, and 99.7% of the data should be in the interval from –13,108 to 41,228 words. These intervals also contained substantially negative numbers. **d)** The data does show that the average number of words for the women in these groups was higher than that for the men; however, the numbers in the intervals were close to each other, and it is not clear whether the differences seen were due to women being more talkative than men or to random differences normally seen when two different samples are taken. In future chapters, we will see how to decide how significant the difference is between groups.

1.99 **a)**

Actual	68	54	92	75	73	98	64	55	80	70
Standardized	–0.2	–1.6	2.2	0.5	0.3	2.8	–0.6	–1.5	1.0	0.0

b) Anyone with a standardized score above 1.04 would receive an A. **c)** The students who received actual total scores of 92 and 98 would receive an A for this course.

1.101 **a)**   Graphs of sales price.

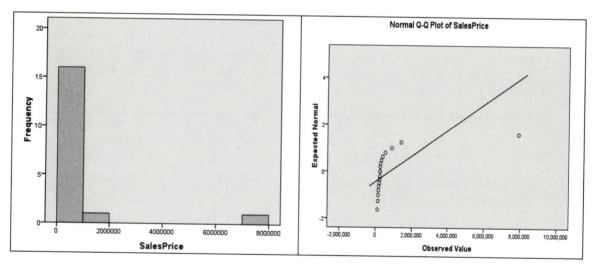

Graphs of building area.

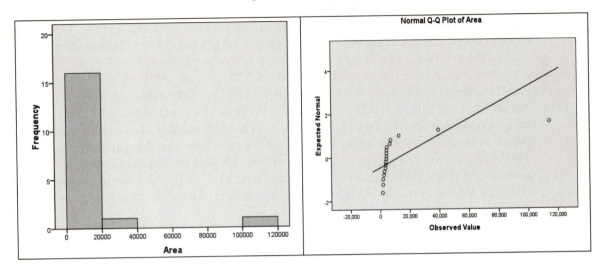

b) Neither graph appears to be Normally distributed. For a Normal distribution, one would expect to see a symmetric histogram. Also, the Normal quantile plots would show that the points followed the forty-five degree angle line. In this case, both of the distributions are skewed to the right with an outlier as seen in both graphs for each variable. **c)** The selling price for the outlier is $7,900,000, and the area is 114,412 square feet.

1.103   **a)** The mean is $65.30 per square foot. The standard deviation is $16.29 per square foot. **b)** The intervals are (49.01, 81.59) for the points within one standard deviation of the mean, (32.72, 97.88) for the points within two standard deviations, and (16.43, 114.17) for the points within three standard deviations. **c)** Tables may vary, but below is one possible presentation.

	Percent of Data between 49.01 and 81.59	Percent of Data between 32.72 and 97.88	Percent of Data between 16.43 and 114.17
If the Data Were Normally Distributed	68%	95%	99.7%
As Seen in the Actual Data	12/18 = 66.7%	17/18 = 94.4%	18/18 = 100%

d) The table does indicate Normality for this set of data.

1.105   **a)** $\bar{x}$ = 579,338; s = 1,496,814. **b)** $\bar{x} \pm 3s = (-3,911,104, 5,069,779)$. **c)** There were no negative volumes in the data, and 21 out of the 22 data points were between 0 and 2,410,204. The data clearly do not follow the 68-95-99.7% rule.

1.107   See the histogram and Normal quantile plot below.

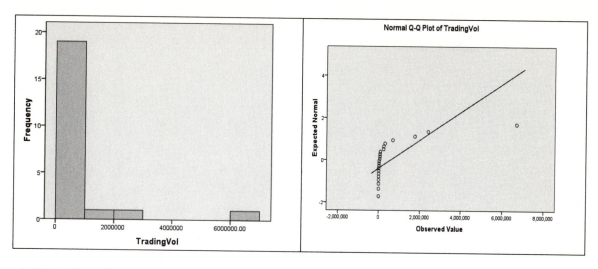

1.109 The taller curve has an approximate standard deviation of 0.2, and the shorter curve has an approximate standard deviation of 0.5. Answers will vary.

1.111 **a)** The probability is 0.0107.

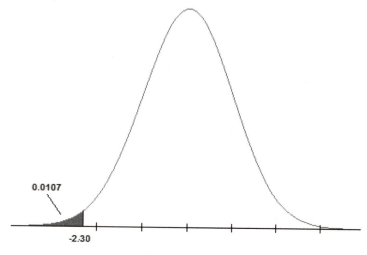

b) The probability is 0.9893.

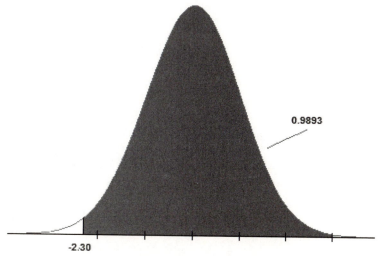

c) The probability is 0.0446.

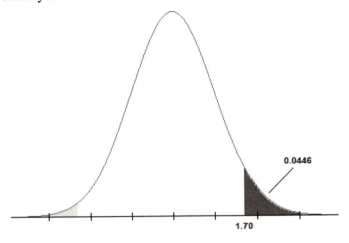

d) The probability is 0.9447.

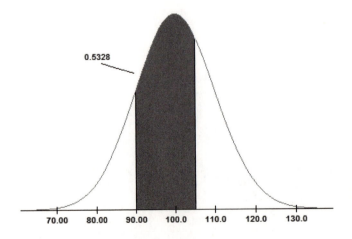

1.113   **a)** The standardized score for 90 is $z = -1.00$. The standardized score for 105 is $z = 0.50$.
The area to the left of $z = 0.50$ is 0.6915. The area to the left of $z = -1.00$ is 0.1587.
Subtracting the two gives an area of 0.5328.

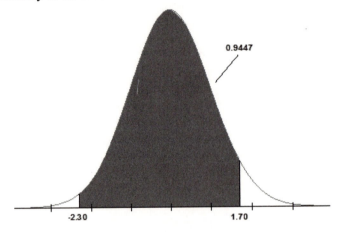

b)

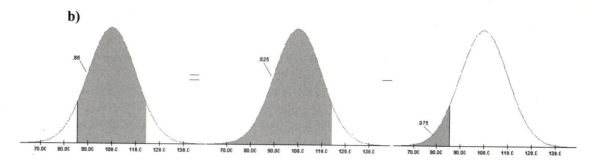

The area between two x values is 85%. The area outside each of these x values is 7.5% on each side. So, we take the area to the left of the upper x value (92.5%) and subtract the area to the right of the lower x value (7.5%) to get the 85%. The z values corresponding to these areas are $z = 1.44$ and $z = -1.44$. The corresponding x values are 85.6 and 114.4.

1.115   **a)** 0.25, $Q_1 = -0.675$, $Q_3 = 0.675$. **b)** $Q_1 = 255.2$, $Q_3 = 276.8$.

1.117   The histograms and Normal quantile plots will differ for each student. Check to see that the student can use the software package correctly.

1.119   (1) goes with picture (c), as there are only two bars and the bar on the right side is somewhat larger. This corresponds to a reasonably larger portion of women in a particular class. (2) goes with picture (b), where there is a greater distinction between the two bars. This corresponds to the understanding that the majority of people are right-handed. (3) goes with picture (d), which shows that there are much smaller numbers of people that are extremely tall or extremely short and that most heights will fall closer to the center of the range. (4) corresponds to (a), showing that the majority of students study for a relatively short time with successively fewer students studying for longer periods of time.

1.121   **a)** The store is located in Iowa. **b)** Since one state has such a large number of stores compared to the others, the pie chart would show one huge slice and many very tiny slices. There would be no way to distinguish between any of the other states except for Iowa. **c)** See the pie chart below.

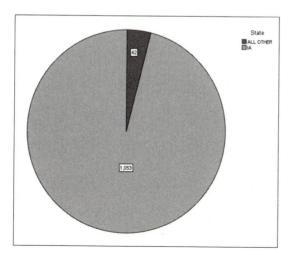

1.123   **a)** The five-number summary is $1.50, $4.80, $6, $10, $78 refunded.
b) 75.73% of all refunds were $10 or less. **c)** The boxplot is shown below.
d) The boxplot shows that the distribution is strongly skewed toward the greater refunds.
Note that this may be skewed left if the data is entered as negative revenue as below or
skewed right if it is entered as dollars refunded.

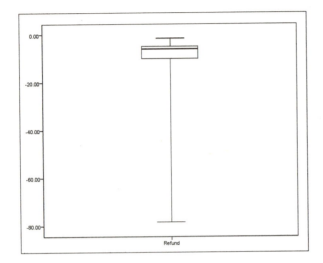

1.125   **a)** $-0.0224 + 3 (0.2180) = 0.6316$. **b)** Since approximately 99.7% of the data should be
within 3 standard deviations of the mean, the portion that should be greater than 3
standard deviations above the mean is 0.15%. **c)** 0%. This percentage is not really
different from the 0.15%, as there are only 22 companies, and 0.15% of 22 cases is 0.033
cases, which rounds to 0%.

1.127   The categorical variables are gender and automobile preference. The quantitative
variables are age and household income.

1.129   The distribution for hatchbacks is skewed to the right with a mean of 22.55 mpg and a
standard deviation of 3.42 mpg. The five-number summary is 16, 20, 21.5, 25, 30.

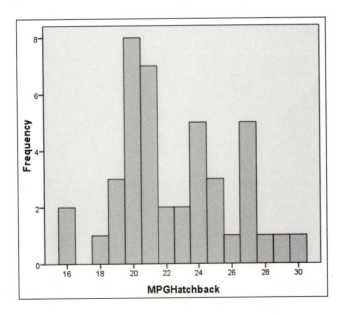

The distribution for large sedans is unimodal and skewed to the left with a mean of 16.57 mpg and a standard deviation of 1.43 mpg. The five-number summary is 13, 16, 17, 17, 19. The large sedan distribution also has two high outliers at 19 mpg and two low outliers at 13 mpg.

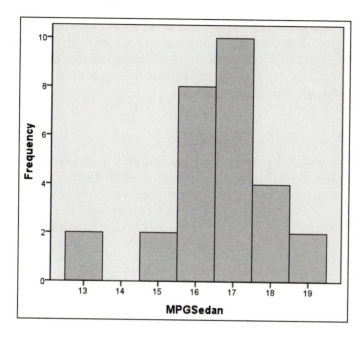

The side-by-side boxplots show that the mpg for hatchbacks is much greater than for large sedans. The range of values for hatchbacks is also greater than the range for large sedans.

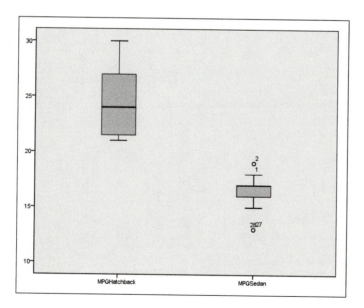

1.131 The IQR is 38.6. The values of 118.2, 156.5, 196, and 204.9 are all high outliers, as they lie more than 1.5*IQR above the third quartile.

1.133 Reports will vary based on year of data selected. Information should be included on the mean, standard deviation and five-number summaries of both the "business starts" and "business failures." The shape of the distribution from either a histogram or a stemplot should be discussed as well as the presence of any outliers. A comparison of the two graphs and a comparison of the numerical summaries for these two variables should also be made.

1.135   Reports will vary based on data selected. Reports should include graphical summary information from histograms, stemplots, and/or boxplots. They should also include numerical summary information such as the five-number summary and/or the mean and standard deviation. The report should mention any outliers present in the data set and should discuss the overall shape of the distribution.

Case Study 1

Bar graphs of the distributions of vehicles by color for the different regions are shown below. Notice that silver is the most common color in all regions. For North America and Japan, white is the second most common color while the remaining regions favor black. Brown is the least common color (or tied for the least common) in all regions. South Korea sticks almost entirely to the seven main colors studied here while 10 to 11% of North American and Chinese cars are colors other than those in the study.

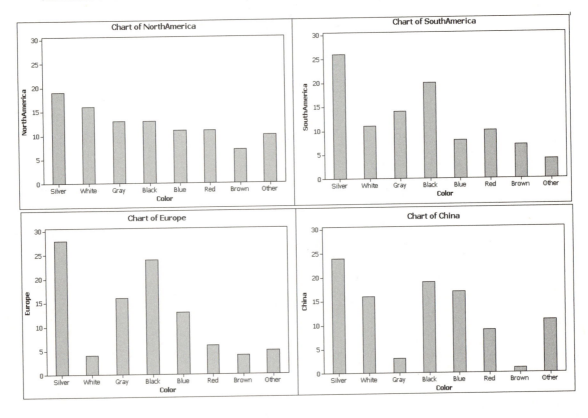

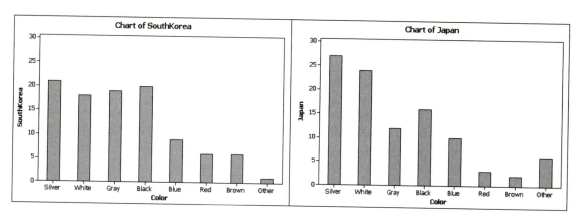

Case Study 2

Answers will vary with factors selected.

Chapter 2: Examining Relationships

2.1 **a)** The cases are individual employees. **b)** The label variable of the data set would be an identifier for the employee. This could be the employee's name or an identifier such as an ID number. **c)** Quantitative variables might include hours of sleep and units produced (or similar measures depending on occupation). Categorical variables might include job type or ranking (manager/worker for example). Other similar answers would be acceptable as well. **d)** There would be an explanatory variable and a response variable if we were interested in trying to show that the number of hours of sleep employees get helps predict how effectively they work.

2.3 **a)**

Botnet	Bots (thousands)	Spams per Day (billions)
Srizbi	315	60
Bobax	185	9
Rustock	150	30
Cutwail	125	16
Storm	85	3
Grum	50	2
Ozdok	35	10
Nucrypt	20	5
Wopla	20	0.6
Spamthru	12	0.35

b) There are 10 cases in the data set; each case is a botnet. **c)** The botnet name is the label. The variables are number of bots in thousands and number of spam e-mails per day in billions. **d)** Both variables listed in part (c) are quantitative variables.

2.5 **a) and b)**

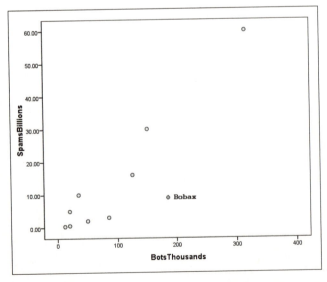

2.7 On average, the amount of debt increased 19.07 million dollars from 2006 to 2007.

2.9   For the lower third of the log unemployment rates, we see that the log GDP per capita is also on the low side. The highest third has log GDP per capita values that are higher in general than the other values. The middle third has values toward the middle of the log GDP per capita range.

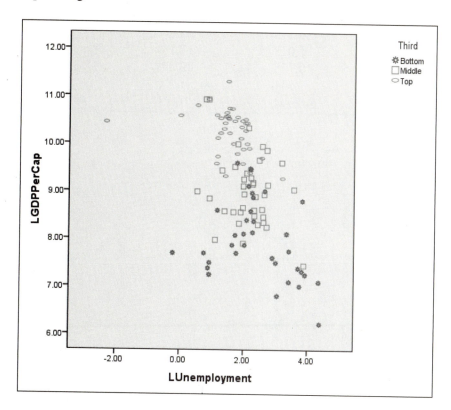

2.11   Answers will vary. Sketches should show that the student understands the difference between strong and weak relationships; between positive and negative relationships; and between linear, non-linear, and no relationship.

2.13   **a)**

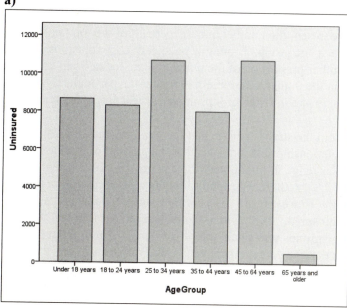

b)

Age Group	Number Uninsured	Percent Uninsured by Age Group
Under 18 years	8,661	18.4%
18–24 years	8,323	17.7%
25–34 years	10,713	22.8%
35–44 years	8,018	17.1%
45–64 years	10,738	22.8%
65 years and older	541	1.2%
Total All Ages	46,994	100%

c)

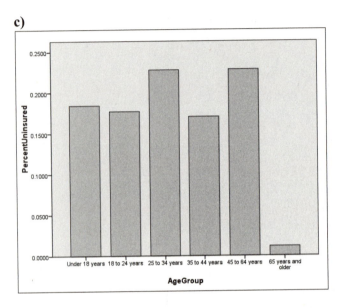

d) The graphs are almost identical except for the scale used. Changing the scale from counts to percents does not change the shape of the graph. **e)** People in the 25 to 34 year old and the 45 to 64 year old groups have the largest number of uninsured. The over 65 age group makes up the smallest percent of the uninsured population.

2.15   In Exercise 2.13, the plots represent the number and the percent of the total uninsured population that are within each age group. From this perspective, conclusions can be drawn about where the greatest number/proportion of the uninsured are located. In Exercise 2.14, the plots show how prevalent the uninsured problem is within the particular age groups. Conclusions can be drawn here about which groups are more/less likely to be uninsured.

2.17   **a)** It cannot be concluded from the strength alone that the numbers remained approximately the same. A strong linear relationship just indicates that the amount of increase/decrease would remain within a similar range throughout all the age groups. **b)** While the number in the population increased by 0.87%, there was approximately a 3% decrease in the number of uninsured in 2007 from 2006. However, the "65 years and older" group saw almost a 27% increase in the number of uninsured. As shown in the bar graph that follows, the amount of change for the individual age groups was small and of similar size throughout the age groups, but chart on the right shows that the percent increase is quite different in the different age groups.

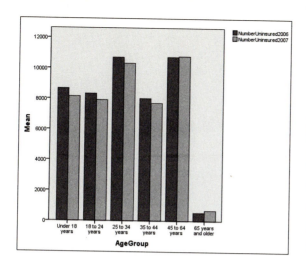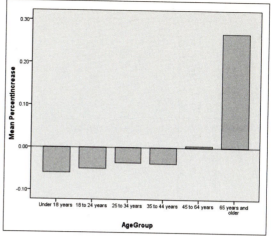

2.19 **a)**

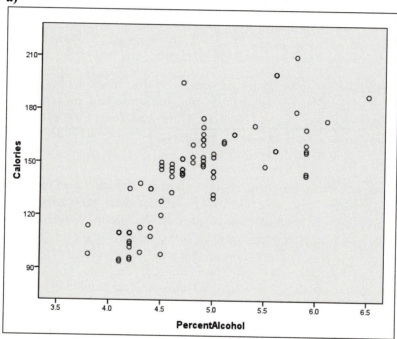

b) The relationship here is a moderately strong, positive, linear relationship, although the relationship is not as strong, nor as obviously linear for the upper values of "percent alcohol."

2.21   **a)**

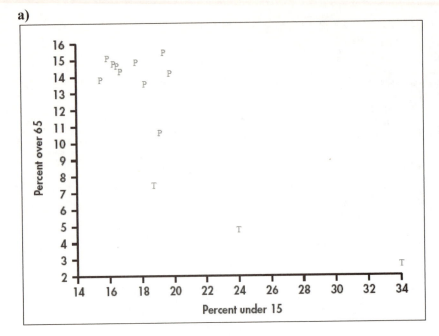

b) For the provinces, the percent over 65 is similar, between 13.5% and 15.5%. The range of percents for those under age 15 varies within these same territories with no particular pattern. However, for the three territories, Yukon Territories, Northwest Territories, and Nunavut, the percent of the population that is over 65 is much smaller and the percent under 15 is much larger.

2.23   The range of life expectancies for European countries is much narrower than the range for all countries. There is a moderately positive relationship between these two variables. Of importance is that there are very few European countries with extremely low Internet usage, and the life expectancy for European countries tends to be higher than for other areas of the world in general.

2.25   The correlation between the number of bots (in thousands) and the number of spam messages (in billions) is $r = 0.884$.

2.27   Since the slope of the regression line is positive and the points on the scatterplot closely follow this line, the correlation is near 1.

2.29   **a)**

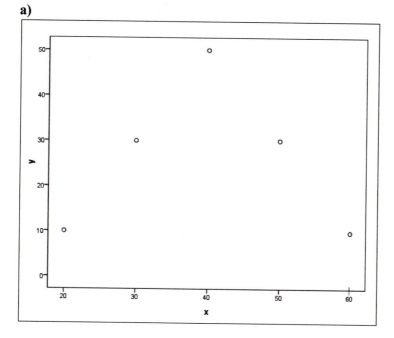

b) The relationship between these points is very strong. The relationship is linear and increasing between the *x*-values of 20 and 40 but linear and decreasing from *x*-values of 40 to 60. **c)** $r = 0$. **d)** Correlation is only an accurate measure of strength when the entire relationship is linear.

2.31   **a)** Correlation is used to describe linear relationships. Since this relationship is not linear, correlation would not be the best measure of the strength of the relationship. **b)** $r = -0.479$. **c)** For log unemployment rates above 1.5, the relationship is linear, so correlation would be a good measure of the strength of the relationship.

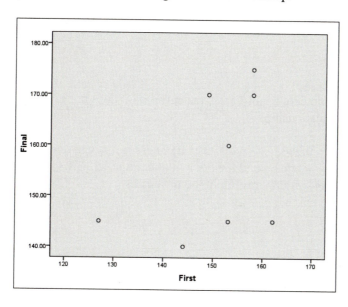

c) $r = 0.409$. **d)** Answers may vary. Some possible reasons include: students who scored low on the first exam may change their study methods or get a tutor to help increase their scores for the remainder of the course, students who scored well on the first exam could be retaking the course but still not understand the later material, students could have pressures outside of class that might affect one or both of these exam scores, etc.

2.33   **a)** As statistics is a course where concepts build upon each other, an understanding of the earlier material is necessary for students to do well on later material. So, students who score well on the second test would likely continue to score well on the final exam and vice versa. **b)** The scatterplot below shows a weak to moderate positive relationship between the second exam score and the final exam score.

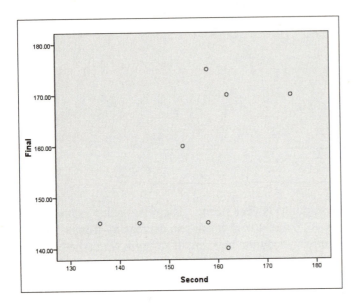

c) $r = 0.519$. **d)** A student who is doing well through the second exam has demonstrated a stronger ability to handle statistical concepts. Students are also likely to have seen some of the information in the beginning of a statistics course in earlier courses, and this understanding may not carry through the remainder of the course. Other answers are also acceptable.

2.35 $r = 0.287$.

2.37 $r = -0.839$. Since the relationship here is not linear, the correlation does not give a good measure of the strength of the relationship.

2.39   **a)** See Exercise 2.21 (a). **b)** $r = -0.967$. **c)** $r = -0.331$. **d)** There is a negative relationship between the percent of people over age 65 and the percent under age 15. This relationship is very strong in the three territories but rather weak in the provinces.

2.41 **a)**

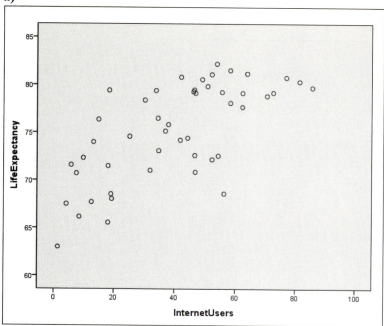

b) $r = 0.686$. **c)** The correlation here is almost the same as that seen when all 181 countries were included; however, the correlation is a better measure to use in this relationship as the data has a relationship that is much closer to a linear form.

2.43 No, unit of measurement changes do not affect the correlation.

2.45 Applet.

2.47 The scatterplot should be similar to that shown below. The correlation is 1.00. When the *y*-variable changes at an exact rate based on a change in the *x*-variable, there is a perfect linear relationship between the two variables. Therefore, the correlation will be 1.00.

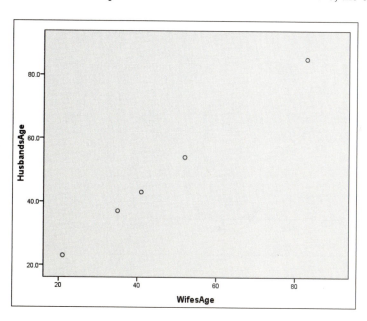

2.49   **a)** There is a moderate, positive linear relationship between these two variables.

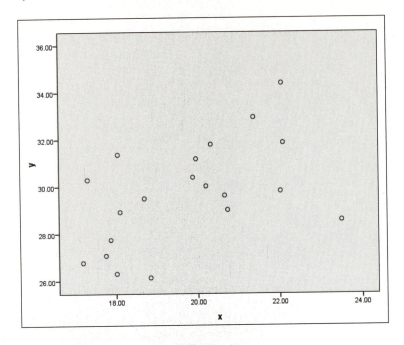

b) $r = 0.523$.

2.51   **a)** In the scatterplot below, we see that the additional data point has a much higher x-value than what was seen in the rest of the points. So, although the y-value is in the range of the remaining points, the x-value makes this point an outlier.

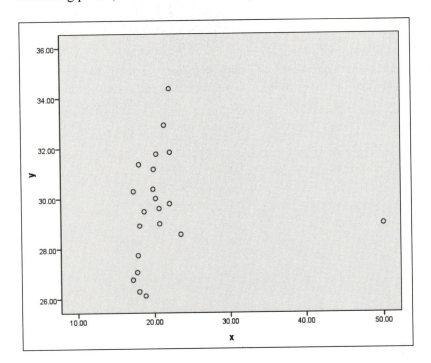

b) $r = 0.072$. This is much smaller than what was seen in the original data set.

c) An outlier that occurs in either the x or y direction alone can cause a relationship to look much weaker than the relationship would appear without the outlier. However, an outlier in both the x and y directions may appear to extend the relationship in the original data set, making it appear stronger than it otherwise would.

2.53 When a line can be drawn through the points in a scatterplot and every point is exactly on that line, the two variables are perfectly correlated. In this case, the equation of the line is $y = 0.5x + b$, where b is the y-intercept of the line, x is Fund A, and y is Fund B.

2.55 $200; $80.

2.57 **a)** The slope of the regression line is 20. **b)** The intercept of the regression line is 10. **c)** When x is 10, the predicted value of y is 210. When $x = 20$, the predicted value of y is 410. When x is 30, the predicted value of y is 610.

d)

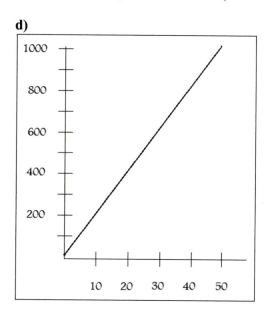

2.59 **a)** 35.5% of the variation in yearly changes is explained by the January change. **b)** The equation of the regression line is $\hat{y} = 6.083 + 1.707x$. **c)** The prediction for the change in the year when the January change is 1.75% is 9.07%. We could have answered this without using the regression equation because we know that the point $(\bar{x}, \bar{y})$ will always fall on the regression line.

2.61 **a)** 28.1196.
b) −19.1196.
c) Bobax has the greater deviation.

2.63   **a)**

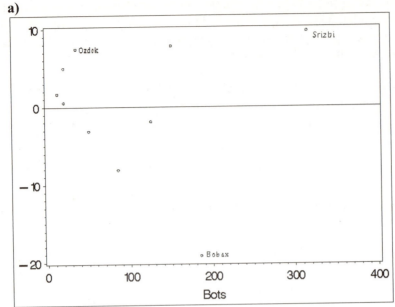

b) These points can be identified by looking for the correct value on the *x*-axis or checking to see that the residual appears to match. Points above the regression line in the scatterplot also have points above the zero line in the residual plot and vice versa.

2.65 $y = 5.00 + 1.10x$.

2.67   **a)** $y = 96 - 6x$.

b)

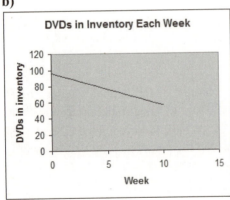

c) No, the initial inventory has only 96 DVDs. This inventory will last for 16 weeks.

2.69 $\hat{y} = 3.379 + 1.615x$.

2.71   **a)** See the plot below. **b)** The equation of the least-squares regression line shown below is $\hat{y} = 318.594 + 0.935x$.

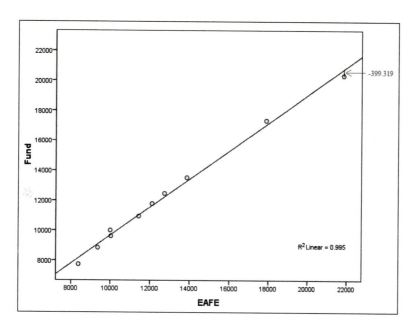

c) The residual using the equation above is –399.319. Answers may vary slightly due to rounding of coefficients by various software programs. SPSS, for example, gives the residual as –408.692. d) The residual plot below shows a steadily decreasing pattern from 1998 to 2002, then sharp increases through 2006 and a sharp decrease again in 2007. This indicates that the fund returns may have a somewhat cyclical pattern compared to their predicted return over time. e) 56.0%.

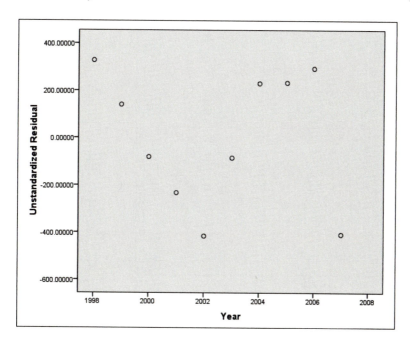

2.73 a) The relationship between y and x is still positive and linear, but the slope of the line showing this relationship is steeper with the addition of the outlier.

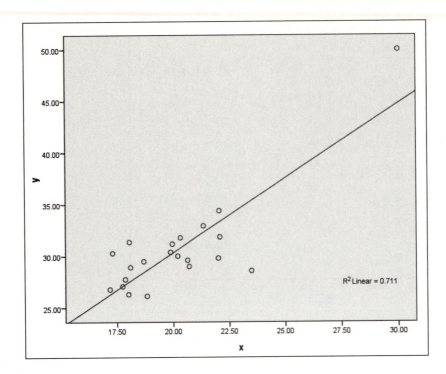

b) $\hat{y} = 1.470 + 1.443x$.

c) The residual plot shows that the new point is definitely an outlier.

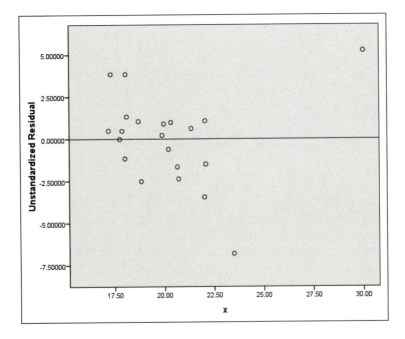

d) 71.1%. **e)** Adding this point to the data causes the relationship between the data points to appear stronger than the initial relationship did. The slope of the regression model is also steeper than with the original model.

2.75

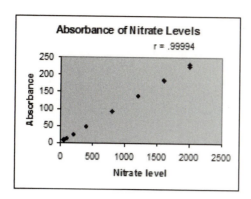

a) $r = 0.9994$, so the calibration does not need to be redone. **b)** The least-squares regression line is $\hat{y} = 1.6571 + 0.1133x$. For a specimen with 500 mg of nitrates, the expected absorbance would be 58.31. The prediction should be quite accurate.

2.77   Applet.

2.79 $r = -0.80$.

2.81   **a)**

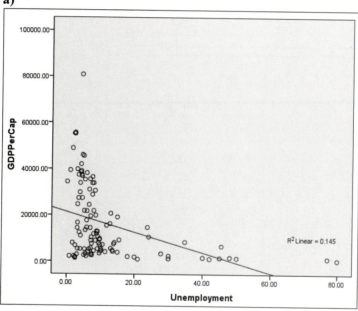

b) A least-squares regression line is not appropriate for describing this relationship as the relationship is not linear. **c)** The residual plot below shows points in an almost v-shaped pattern, indicating that there is quite a bit of variation in the residuals at the lower unemployment values with only a small number of points extending to the right and the higher unemployment values. When the relationship between the variables is linear, the plot should show a random scattering of points without any pattern to the residuals.

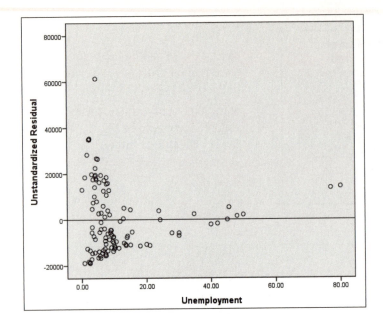

2.83   **a)**

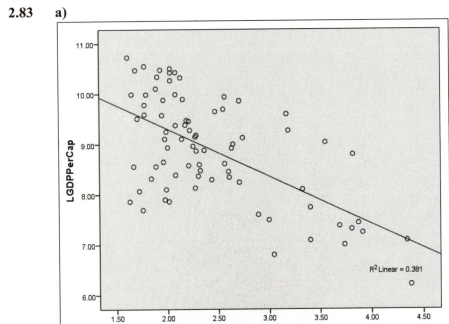

b) Since the relationship here is linear, it is appropriate to use the least-squares regression line to describe this relationship. **c)** The plot that follows shows rather random scatter around the residual line with a larger concentration of points toward the left side, indicating that the lower log unemployment rates are more common than the higher ones.

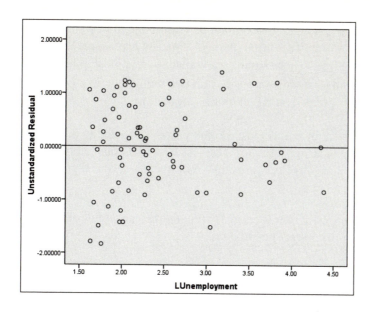

2.85

City	Residual
Los Angeles	5,994.828
Washington, D.C.	2,763.743
Minneapolis	2,107.588
Philadelphia	169.413
Oakland	27.908
Boston	20.958
San Francisco	−75.782
Baltimore	−131.552
New York	−283.037
Long Beach	−1,181.71
Miami	−2,129.21
Chicago	−7,283.28

2.87 **a)**

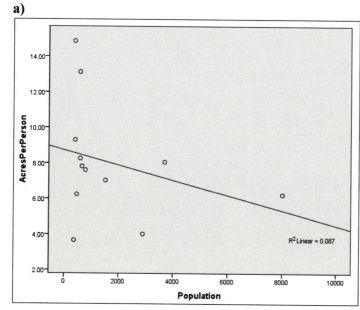

b) The relationship here is not linear, so it would not be reasonable to fit a straight line to the data. (The line is shown anyway in response to part (c).) **c)** The least-squares regression line is $\hat{y} = 8.739 - 0.00042x$. **d)** 8.7%. These results show that the relationship here is not linear as it was in Exercise 2.84. The least-squares regression line would not do a good job of predicting open space in acres per person based on the population.

2.89   The sketch should show a cluster of points closer to the *x*-axis representing academia and a cluster of points closer to the *y*-axis representing businesses so that connecting these two clusters with a line results in a negative relationship. Within each cluster there should be a strong positive relationship.

2.91   No, larger hospitals most likely treat more seriously ill patients than smaller hospitals. Seriously ill patients tend to have longer hospital stays.

2.93   **a)** When all points on the scatterplot are above a negatively sloped regression line, the residuals would be positive but the relationship between the variables would be negative. **b)** A negative or positive relationship between the variables describes how the variables are related, not why they are related. When decreases in the explanatory variable cause increases in the response variable, a negative causation relationship would exist. **c)** Lurking variables help explain the relationship between the explanatory and the response variables. They are not the response variables themselves.

2.95   No. Countries that have high Internet usage are likely to be wealthier countries. These countries would also have better access to medicine, health care, and healthy foods.

2.97   Answers will vary but may include age, education, years of experience in job, and location.

2.99   A more plausible explanation is that heavier people tend to be on diets and try to reduce calories from sugar. Therefore, they often use artificial sweeteners.

2.101 A possible lurking variable could be that higher paying jobs carry with them an expectation that the employee will continue his education while employed. Also, social status may play a role. Affluent families can afford to pay for higher education. The men from these families also have access to higher paying jobs through their contacts.

2.103

	Count	Percent
None	84	34.6%
French	75	30.9%
Italian	84	34.6%

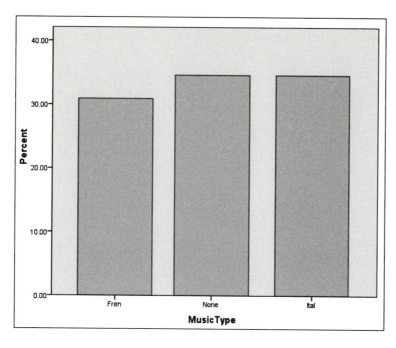

2.105 **a)**

Field of Study	Canada	France	Germany	Italy	Japan	U.K.	U.S.	All
Social sciences, business, law	64	153	66	125	259	152	878	1,697
Science, mathematics, engineering	35	111	66	80	136	128	355	911
Arts and humanities	27	74	33	42	123	105	397	801
Education	20	45	18	16	39	14	167	319
Other	30	289	35	58	97	76	272	857
Total	176	672	218	321	654	475	2,069	4,585

b)

Canada	France	Germany	Italy	Japan	U.K.	U.S.
3.8%	14.7%	4.8%	7.0%	14.3%	10.4%	45.1%

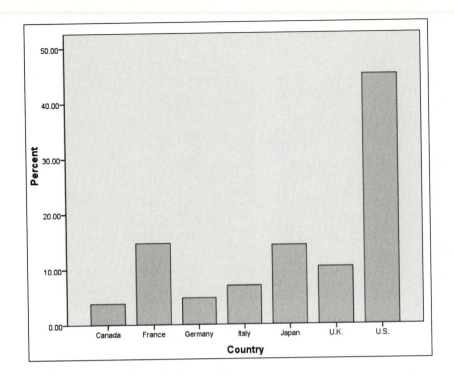

c)

Social sciences, business, law	37.0%
Science, mathematics, engineering	19.9%
Arts and humanities	17.5%
Education	7.0%
Other	18.7%

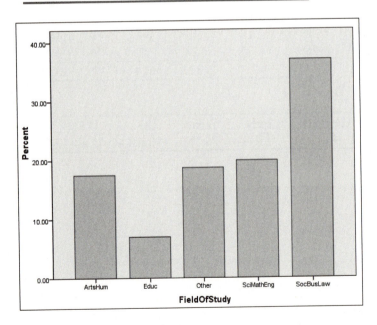

2.107 **a)**

	Italian Music
French Wine	30
Italian Wine	19
Other Wine	35
Total	84

b)

	Italian Music
French Wine	0.357
Italian Wine	0.226
Other Wine	0.417

c)

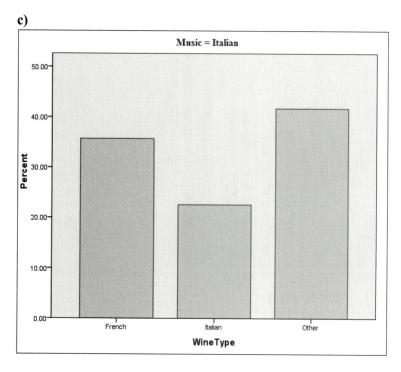

d) There were more bottles of Italian wine sold when Italian music was playing than when no music was playing.

2.109 **a)** The chart below shows the conditional distributions of field of study for each of the countries.

Field of Study	Canada	France	Germany	Italy	Japan	U.K.	U.S.
Social sciences, business, law	36.4%	22.8%	30.3%	38.9%	39.6%	32.0%	42.4%
Science, math, engineering	19.9%	16.5%	30.3%	24.9%	20.8%	26.9%	17.2%
Arts and humanities	15.3%	11.0%	15.1%	13.1%	18.8%	22.1%	19.2%
Education	11.4%	6.7%	8.3%	5.0%	6.0%	2.9%	8.1%
Other	17.0%	43.0%	16.1%	18.1%	14.8%	16.0%	13.1%

b)

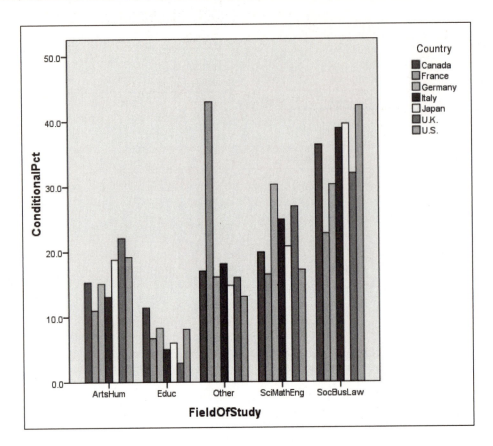

c) The bar graph shows that most countries have a greater percent of their students in social sciences, business, and law. The exception to this is France, where "other" is the most common field of study. In most countries, science/math/engineering is the second most common field of study; however, in the United States, arts and humanities is the second most common choice. Education has the smallest percent for all of the countries.

2.111 **a)** These approaches allow us to study different things (see part (c)). There are some items that stand out in both cases, such as the large number of students in France in "other" areas of study. **b)** When looking at the conditional distribution by field of study, we are primarily seeing the difference in the sizes of college populations between the countries. If we are interested in studying the popularity of the different fields of study, the conditional distributions of field of study given the different countries would be more instructive. **c)** The first of these approaches allows us to see which fields are most/least popular across the different countries, while the second approach allows us to see where the greatest number of students are within each field.

2.113 **a)** 3.8%, 4%. Poor patients fare better in hospital A. **b)** 1.0%, 1.3%. Good patients fare better in hospital A also. **c)** Based on parts (a) and (b) alone, I would recommend that someone facing surgery have the surgery completed in Hospital A. However, only 2% of the overall total number of surgeries result in deaths in Hospital B, whereas 3% of those in Hospital A result in deaths. **d)** The number of people who are in "poor" condition prior to surgery is much greater at Hospital A, and the percent of those in "poor" condition who survive after the surgery is less than the percent of those in "good" condition who survive.

2.115 The table below gives the percent of banks offering RDC within the given regions.

Region	Northeast	Southeast	Central	Midwest	Southwest	West
% with RDC	42.4%	48.3%	38.7%	25.8%	34.6%	44.5%

This bar graph shows the percent within each region that offer RDC.

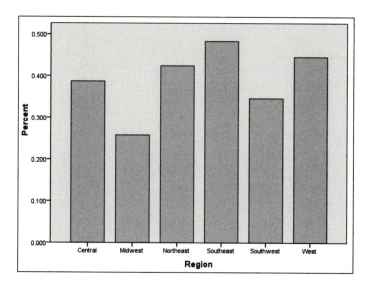

The following bar graph below shows the number of banks in each region.

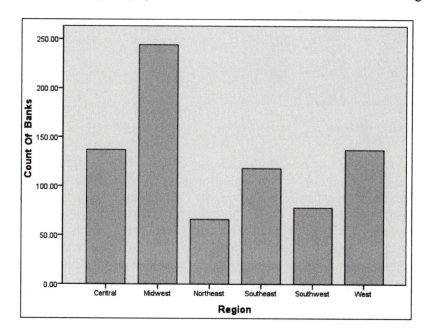

From the two bar graphs above, we can see that while the Midwest has a larger number of banks than any other region, they have the smallest percent of banks with RDC. In contrast, the Northeast has the smallest number of banks but one of the largest percents of banks with RDC. The southeast has the largest percent of banks with RDC.

2.117 **a)** High exercise—56.77%, low exercise—43.23%. **b)** High exercise—37.95%, low exercise—62.05%. **c)** Those who get an adequate amount of sleep tend to be in the high exercise group while those who do not get an adequate amount of sleep tend to be in the low exercise group. **d)** It is easier to see that higher levels of exercise lead to adequate sleep than to see that adequate sleep leads to higher exercise levels, so the summary from Exercise 2.116 seems preferable.

2.119 a)

Age Group	Full Time	Part Time
15–19	89.7%	10.3%
20–24	81.8%	18.2%
25–34	50.1%	49.9%
35 and over	27.1%	72.9%

b)

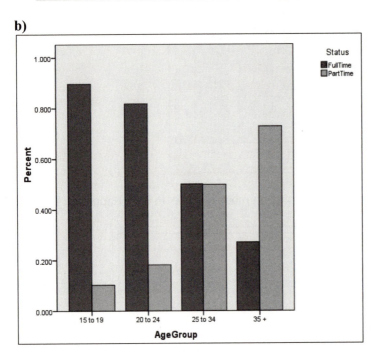

c) Students aged 15 to 24 are more likely to be full-time students than part-time students. Students age 25 to 34 are equally as likely to be full-time or part-time. Students age 35 and older are much more likely to be part time students. **d)** The full-time and part-time percents add to 1 in each age group, so knowing one of these also gives knowledge of the other. **e)** Using only the summary in Exercise 2.118, we do not see the impact of the sizes of the various age groups. For example, while the students ages 35 and over make up 38.6% of part-time students, this is even more interesting when we realize that this group was the smallest group to begin with and that almost 73% of this group are part-time students.

2.121 a)

	Male	Female	Total
Agree	11,724	14,169	25,893
Disagree	1,163	746	1,909
Total	12,887	14,915	27,802

b) 93% of students agreed that honesty and trust were essential in business and the workplace, and 7% disagreed. The percentages were very similar regardless of gender. 91% of males and 95% of females agreed, while 9% of males and 5% of females disagreed.

c) When asked whether students agreed or disagreed about whether trust and honesty were essential in the workplace, most (93%) of them agreed that they were. A slightly smaller proportion of males (91%) agreed with this statement than females (95%).

2.123   **a)** *P*(Hired | age < 40) = 0.0639, *P*(Hired | age ≥ 40) = 0.0060.

b)

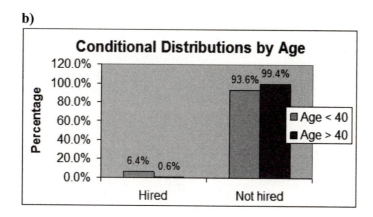

c) A greater percent of applicants in the under-40 age group are hired as compared to those applicants who are 40 or older. **d)** A lurking variable could be whether applicants are qualified for the position or not.

2.125   **a)** 25,262 were never married; 65,128 were married; 11,208 were widowed; 13,208 were divorced. **b)** The joint distribution is below.

Age	Never Married	Married	Widowed	Divorced
18–24	9.9%	2.1%	0.0%	0.2%
25–39	7.3%	16.8%	0.1%	2.2%
40–64	4.2%	29.8%	2.0%	7.4%
≥ 65	0.7%	8.0%	7.6%	1.7%

The marginal distribution of marital status is 22.0% never married, 56.7% married, 9.8% widowed, 11.5% divorced. The marginal distribution of age is 12.2% aged 18 to 24, 26.4% aged 25 to 39, 43.5% aged 40 to 64, and 17.9% aged 65 and over.

Each row of the following table shows the conditional distribution for a given age.

Age	Never Married	Married	Widowed	Divorced
18–24	81.3%	17.3%	0.1%	1.3%
25–39	27.5%	63.7%	0.6%	8.3%
40–64	9.6%	68.6%	4.7%	17.1%
≥ 65	3.7%	44.5%	42.2%	9.6%

Each column of the following table shows the conditional distribution for a given marital status.

Age	Never Married	Married	Widowed	Divorced
18–24	45.0%	3.7%	0.2%	1.4%
25–39	33.0%	29.7%	1.5%	19.0%
40–64	19.0%	52.6%	20.8%	64.6%
≥ 65	3.0%	14.1%	77.5%	15.0%

c) While the "married" status shows the greatest percent of all women, this is not true for all age ranges. For the 18 to 24 age group, "never married" has the greatest percent. Likewise, the percent who are widowed is much greater for the over 65 age group than any of the other age groups. The largest age group is women aged 40 to 64. These women also make up the greatest percent of divorced women and the greatest percent of married women.

2.127 **a)** 30,856 were never married, 64,590 were married, 2693 were widowed, and 9610 were divorced. **b)** The joint distribution is below.

Age	Never Married	Married	Widowed	Divorced
18–24	12.0%	1.2%	0.0%	0.1%
25–39	10.7%	15.8%	0.1%	1.7%
40–64	5.4%	32.1%	0.6%	6.1%
≥ 65	0.6%	10.8%	1.9%	1.1%

The marginal distribution for marital status is: 28.6% never married, 59.9% married, 2.5% widowed, and 8.9% divorced.

The marginal distribution of age is 13.4% aged 18 to 24, 28.2% aged 25 to 39, 44.1% aged 40 to 64, and 14.3% aged 65 and over.

Each row of the following table shows the conditional distribution for a given age.

Age	Never Married	Married	Widowed	Divorced
18–24	89.9%	9.3%	0.1%	0.7%
25–39	37.8%	56.0%	0.2%	6.0%
40–64	12.2%	72.8%	1.3%	13.8%
≥ 65	4.0%	75.4%	13.1%	7.5%

Each column of the following table shows the conditional distribution for a given marital status.

Age	Never Married	Married	Widowed	Divorced
18–24	42.0%	2.1%	0.4%	1.0%
25–39	37.3%	26.4%	2.0%	18.9%
40–64	18.7%	53.5%	22.7%	68.0%
≥ 65	2.0%	18.0%	74.9%	12.1%

c) The largest group of never-married men is the 18 to 24 age group. Those who are married tend to be in the 25 to 64 age range. Men who are widowed are mostly ages 40 to 64 or over 65, while divorced men are primarily ages 40 to 64. More than half of men who are married are 25 years old or above, and over ¾ of men over 65 are married. Most 18- to 24-year-old men have never married (89.9%). The largest single segment of this population is men aged 40 to 64 who are married (32.1% of the population).

2.129   Answers may vary. Here is one possible set of tables that will illustrate Simpson' paradox.

Smokers

	Overweight	Not
Early Death	10	250
No	200	550

Nonsmokers

	Overweight	Not
Early Death	300	70
No	500	1200

Grouped Together

	Overweight	Not
Early Death	40	950
No	700	1750

2.131   **a)** The scatterplot below shows a rather weak, positive linear relationship between dwelling permits and sales. There do not appear to be any outliers or influential observations in this data set.

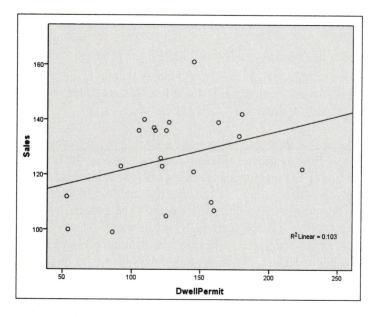

b) The least-squares regression line seen above is $\hat{y} = 109.82 + 0.126x$.

c) 129.98. **d)** The residual is −22.98. **e)** 10.3%.

2.133   **a)** The scatterplot below shows a weak, positive linear relationship between production and sales. There do not appear to be any outliers or influential observations in this data set.

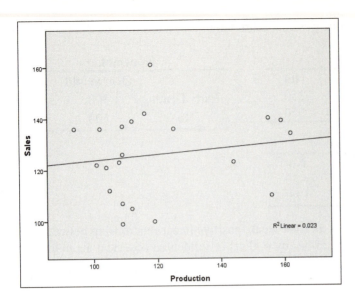

b) The least squares regression line seen above is $\hat{y} = 112.257 + 0.115x$.

c) 126.632. **d)** Finland has actual sales of 136, so the residual is 9.368. **e)** 2.3%. This is only slightly larger than in the previous problem, but still much less than that seen in the relationship between dwelling permits and sales.

2.135 **a)** We see from the residual plot that there is a pattern to the residuals. In the early years, the points had positive residuals, indicating greater salaries than predicted by the linear model. In the middle years (years 6 to 16), the residuals were negative, indicating a smaller salary than predicted by the model. The later years show positive residuals again, indicating that the salary is increasing faster at this point than during earlier years. The pattern to the residuals indicates that the data does not follow a linear pattern. **b)** While we could see that there was a slight deviation from the linear pattern in the scatterplot, the residual plot magnifies this fact, making it more obvious that the salary is not maintaining a linear relationship throughout the years.

2.137 **a)** The regression equation is $\hat{y} = 41.253 + 3.933x$. The predicted salary would be $139,579.

b) The regression equation for predicted log salary is $\hat{y} = 3.868 + 0.048x$. This results in a log salary of $15,868 or an actual salary of $158,856. **c)** The results from the log salary prediction will be better as this relationship followed a linear pattern and the original salary prediction model did not. **d)** While the predictions will be relatively close for both models within the range of years studied, they will not produce similar predictions outside this range. In the original salary model, the salaries were increasing faster than the linear model in the upper years. Predictions should take this increase into account, and only the log salary model does this. **e)** Numerical summaries do not tell the whole story of the data. Using graphical information helps us to see what is really going on. Trying to extrapolate using just the information gained from the numerical summaries would lead to incorrect conclusions when there is not complete understanding of the pattern in the underlying data.

2.139 **a)** $\hat{y} = 5403 + 0.982x$. **b)** There are no apparent outliers in the residual plot below. There is also no particular pattern to the residuals.

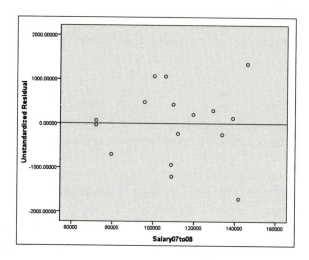

2.141 The raw graduation rate would tell the percent that graduate, but the residuals will tell whether this percent is low or high compared to the predicted rate. A school with a large positive residual has a better than expected graduation rate, while a school with a negative residual has a graduation rate that is lower than predicted.

2.143 **a)** The scatterplot below shows a moderately strong, positive, linear relationship between the proportion of adults who have completed college and the proportion who eat at least five servings of fruits and vegetables a day. **b)** The line below does show the general pattern of the data although there does appear to be a slight U-shaped pattern.

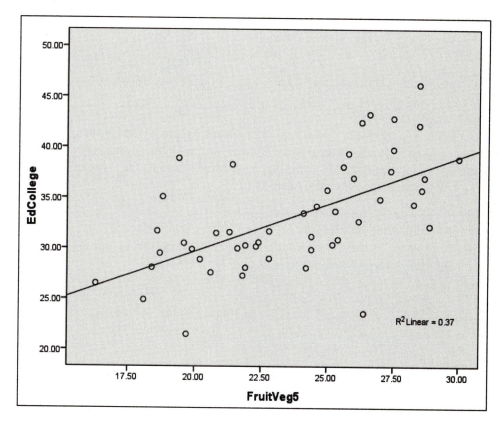

c) Answers will vary depending on states selected. **d)** This relationship only shows that those who earn a college degree also tend to eat at least five servings of fruits and vegetables each day, it does not show that this relationship is due to a causation response.

2.145   Answers will vary based on team chosen and season.

2.147   If the relationship between diversity and population variation is linear, this information supports a moderately strong negative relationship. So as diversity increases, variation in population size decreases.

2.149   **a)** The relationship in the problem suggests that the amps of the saw are helpful in predicting the weight of the saw, so amps should be the explanatory variable. **b)** See the scatterplot below. There is a moderately weak, positive, linear relationship between weight and amp values. **c)** $\hat{y} = 5.8 + 0.4x$, $s = 0.782$, $r^2 = 45.7\%$. **d)** For every 1-amp increase, there is a 0.4-pound increase in weight, on average. **e)** A 2.5-amp increase would correspond to a 1-pound increase in weight, on average. **f)** See the residual plot below. There seems to be a general upward trend, but it is hard to determine if there is curvature.

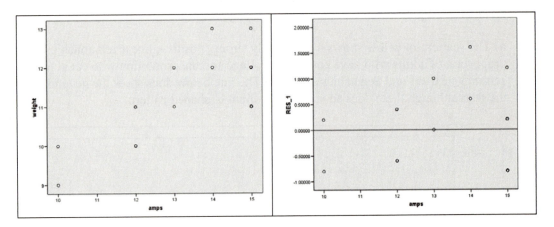

2.151   For two funds to be perfectly correlated, their returns simply have to be proportional to each other. For every $1 fund A returns, fund B might return $0.50. Using fund A as the explanatory variable and fund B as the response variable means the slope of the line is less than 1. The graph below illustrates this point.

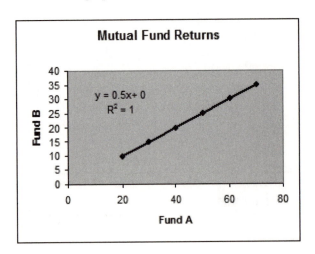

2.153 **a)**

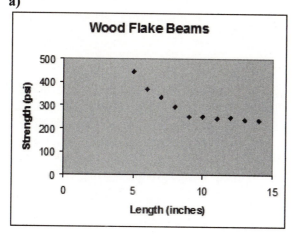

b) The overall pattern is nonlinear. There do not appear to be any outliers.
c) No, a straight line is not a good fit, even though the *r*-squared value might suggest otherwise. The scatterplot clearly shows a nonlinear pattern.

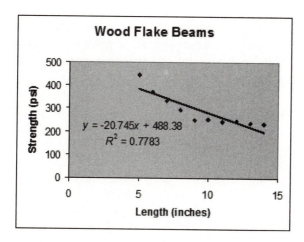

d) These two lines show a much better representation of the data. I would ask the experts why the strength levels out after 9 inches.

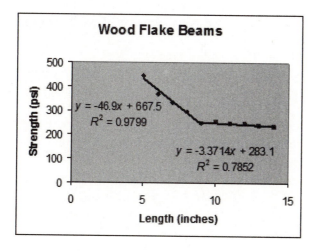

2.155 a)

	Smoker	Not
Dead	139	230
Alive	443	502

The percentage of smokers that were alive after 20 years is 76%, and the percentage of nonsmokers that were still alive after 20 years is 69%. **b)** Within the age group 18 to 44, 93% of the smokers were alive after 20 years and 96% of the nonsmokers were still alive. Within the age group 45 to 64, the percentages were 68% and 74%, respectively. Within the age group 65 and older, the percentages were 14% and 15%, respectively. **c)** This explanation makes sense. The percentage of smokers in the three age groups, starting youngest to oldest, are 46%, 55%, and 20%.

Case Study 2.1

This report contains many graphical and numerical summaries regarding beef consumption in the United States. Students may comment on many items in this 25-page report. Here are a few of the highlights.

Ground beef makes up the largest portion of beef consumed (42%), followed by steak (20%) and processed beef and stew meat (13% each). A greater number of pounds of beef per capita are consumed by lower-income consumers than by higher-income consumers. Ground beef and processed beef use decreases as income increases.

Consumers who are black, non-Hispanic eat a greater number of pounds of beef per capita than other races. Ground beef makes up the greatest portion of pounds consumed regardless of race. Black, non-Hispanic consumers and Hispanic consumers eat a greater number of pounds of steak per capita than white, non-Hispanic consumers or those of other races. Processed beef finds a greater market among the black, non-Hispanic consumers.

65% of all beef consumed is assumed to be consumed at home compared to away from home or in restaurants. Again, ground beef is the leader regardless of the location of consumption; however, the difference in cut of meat is much greater for those eating away from home or in restaurants, where ground beef makes up the vast majority of beef consumed. The lower the income of the consumer, the more likely they were to eat beef at home instead of elsewhere. Studying regions of the country resulted in a larger consumption of beef in the Midwest than in other regions; however, the Midwest had the lowest consumption of steak per capita of all of the regions.

Consumers in rural regions consumed more beef per capita than consumers in suburban or urban regions. The largest differences were in consumption of ground beef and stew meat.

Men in the 20 to 39 age range eat more beef per capita than any other group. Men in general eat a considerably higher amount of beef than women for all age groups. Among women, those in the 12 to 19 age group eat more meat than the other age groups, particularly of ground beef.

Case Study 2.2

A.

	Correlation with GPA
HSM	0.436499
HSS	0.329425
HSE	0.289001
SATM	0.251714
SATV	0.11449

The correlation values for the explanatory variables tell us the strength of the linear relationship between each variable and GPA. HSM and HSS are the two best predictor variables. This makes sense for computer science majors.

B.

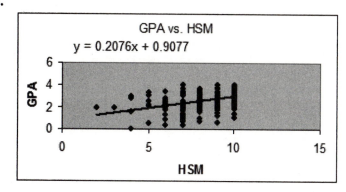

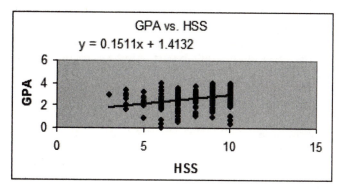

The scatterplots show that, while there is a relationship between HSS and GPA and HSM and GPA, there are other variables that would probably help the prediction. There is one unusual observation seen on the plot of GPA versus HSS. This point shows a fairly high GPA, but the student had a low HSS score.

Case Study 2.3

A.

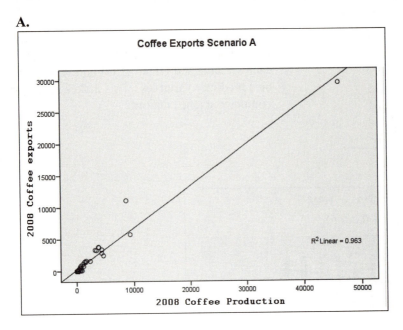

Brazil has extremely high coffee production and exports compared to the rest of the countries in the data set. Indonesia and Columbia also have much higher production and exports than the other countries. Columbia has exports that are much higher than would be predicted based on the amount of production for this country. This can be seen on the scatterplot above. The correlation between 2008 exports and 2008 production is very strong with an $r = 0.981$. Approximately 96.3% of the variation in 2008 exports can be explained by the 2008 production. The prediction equation is given by $\hat{y} = 208.58 + 0.659x$.

B.

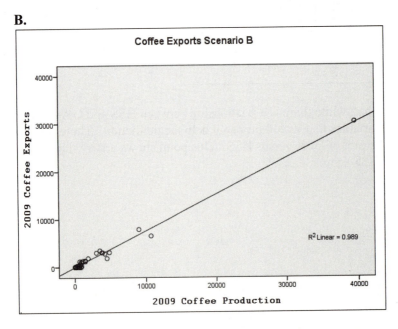

The correlation between 2009 coffee production and 2009 coffee exports is $r = 0.995$. Approximately 98.9% of the variation in 2009 exports can be explained by 2009 production.

The three countries mentioned in part (A) still have very high export values but Columbia's exports are now more in line with the predicted exports based on production. The regression equation is $\hat{y} = -28.896 + 0.763x$.

C.

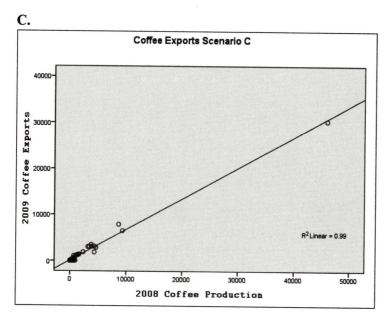

The correlation between 2008 coffee production and 2009 coffee exports is 0.995. Approximately 99% of the variation in 2009 coffee exports can be explained by 2008 coffee production. The regression equation is $\hat{y} = -102.926 + 0.668x$.

D. Below are the scatterplots of scenarios A, B, and C with the three outliers removed. Linear functions do not do the best job of explaining the results here. Transformations may be appropriate using cubic functions or the relationships between the natural logs of both variables.

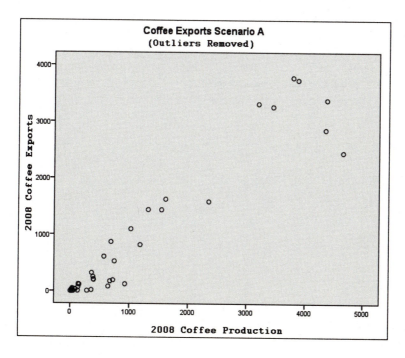

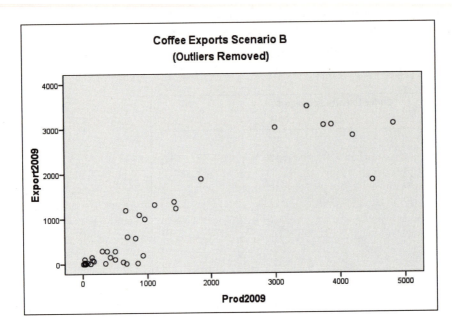

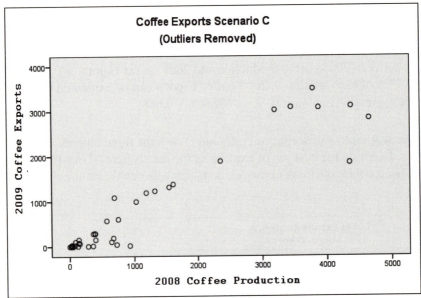

Chapter 3: Producing Data

3.1   The interest of a small group of people can not be used to draw conclusions about the general interest of the population. This is anecdotal evidence.

3.3   Ashley's preference could be due partly to the fact that she is a runner. Her preference would not necessarily generalize to all young people. This is anecdotal evidence.

3.5   Answers will vary depending on the data collected.

3.7   Since the treatments, injection of a vaccine or no injection, were imposed on the adults, this is an experiment rather than an observational study. The explanatory variable is whether the adult received the vaccine. Response variables were the presence of the various adverse affects and whether there was a presence of the antibodies in the adult. Other variables may also have been recorded.

3.9   Since no treatment is imposed and information is just gathered on the consumers, this is an observational study. The explanatory variable is gender, and the response variable is the health plan chosen.

3.11 Certainly the state of the economy will affect the unemployment rate. The change in population over time with respect to demographics and size will also affect the unemployment rate. The efficiency of the workers may also have changed with the implementation of the program causing fewer workers to be needed for these jobs.

3.13 Other information required: whether the 772 forest owners constitute the population of all forest owners or whether this is just a sample. The population would include all forest owners from this region. If the survey was mailed to only some of the forest owners, then the sample includes the 772 forest owners who were asked to complete the survey. The response rate is 45.1%.

3.15 **a)** The population consists of all U. S. adult residents. **b)** The population is made up of all U. S. households. **c)** The population would include all regulators from the supplier *or* just the regulators in the last shipment.

3.17 Answers will vary.

3.19 The sample of junior associates is 29, 07, 22, and 10. Senior associates would be 05 and 09. Names may vary based on assignment of numbers. As an alternative method, the senior associates may be numbered from 0 through 9. If this is done, the senior associates selected would be numbers 2 and 5.

3.21 **a)** Households who do not own phones or who are unlisted will not be included (examples may vary). **b)** Those people with unlisted numbers will be included in the sampling frame.

3.23 **a)** People tend to report their successes and not their failures. The reported results may not be a good reflection of the actual results. There also could be an issue with the playing ability of the opponent. To make this an experiment, randomly select a number of people to play against the friend. The actual outcomes should be recorded.

b) Students who sit in the first two rows may also be students who are interested in the material being taught or who are more determined to do well in the class. To make this an experiment, students would be randomly assigned to a particular seat from the first day of class. Average scores on the first exam would then be compared by rows.

3.25   **a)** This method does not involve any random selection. Reading levels will not necessarily be the same throughout a textbook. To determine the reading level, all passages in the text should be numbered. Passages to be tested would then be randomly selected using a random number table or random selection with a computer application. **b)** Students who attend a 7:30 class may be different types of students than the general population. To randomly select students, a list of all students should be used, and the students should be randomly selected from this list to participate. **c)** Subjects with names toward the end of the alphabet do not have a chance of being selected to be in the population. To do the randomization correctly, a list of all subjects should be used and the 10 subjects should be randomly selected from this list.

3.27   **a)** The population was all U.S. adults aged 18. The sample size was 1001. **b)** Including all of the 1001 respondents gives a better overall view of the popularity and visibility of the person. A person will not be popular with a group that has never heard of them. However, in deciding whether a news personality does a good job or not, a survey of just those who have heard of the person will give a better indication of the quality of their work. Preference for method would depend on the purpose for the study and may vary.

3.29   Starting at line 137, the sample numbers are 12, 14, 11, 16, and 08. The apartment complexes will vary depending on numbering scheme. A possible result is Country View, Crestview, Country Squire, Fairington, and Burberry.

3.31   Applet. Answers will vary.

3.33   Number the blocks from 01 to 44, where 01 corresponds to block 1000 and 44 corresponds to block 3025. Starting at line 115 in Table B, the sample numbers are 04 (block 1003), 17 (block 2010), 32 (block 3013), 22 (block 3003), and 09 (block 2002).

3.35   Start by numbering the blocks 1000 to 1005 from 1 to 6 in order. Using Table B, select the block corresponding to the first number (from 1 to 6) that appears in the selected row. Then number the blocks in Group 2 from 01 to 12. Using a new line on the table or starting from the place where you stopped after Group 1, select the first two of the numbers from 01 to 12 that appear in this row. Finally, number the Group 3 blocks from 01 to 26. Using a new line or starting from the place where you stopped after Group 2, select the first three numbers from 01 to 26 that appear. Answers will vary depending the line(s) of the table used.

3.37   Every number from 1 to 45 has an equal an equal chance of being chosen (1/45). Each number from 46 through 90 will be selected based on the initial number selected. So, for example, number 46 will be chosen every time number 1 is chosen. Thus, number 46 has a 1/45 chance of being selected. Likewise, every number between 91 and 135 has a 1/45 chance of being selected and every number between 136 and 180 has a 1/45 chance of being selected. So, each number from 1 to 9000 has a 1 in 45 chance (equal chance) of being selected. However, each set of 4 individuals does not have an equal chance of being selected. For example, numbers 1, 2, 3, and 4 will not ever be selected as the sample. Only the 45 combinations discussed will have a chance.

3.39 Number the parcels in Climax 1 with the numbers 01 through 36. Number the parcels in Climax 2 with the numbers 01 through 72. Number the parcels in Climax 3 from 01 through 42. Starting at line 130 in Table B, select the parcels from Climax 1. These would be numbers 05, 16, 17, and 20. Then continue immediately from the place where the last number in this sample was selected and select the parcels from Climax 2. The parcels selected would be numbers 19, 72, 45, 05, 71, 66, and 32. Continuing on, select the parcels for Climax 3. These would be 19, 04, and 25. Finally, continue on immediately after the last selected number and select the Secondary parcels. These would be 29, 20, 16, and 37. Other selection methods may be used as well, such as numbering parcels starting at 00 instead of 01, numbering the parcels from 001 to 181, only looking for the numbers within the forest type you are working on at any time, or starting with a new line after each type is completed.

3.41 **a)** This is a poorly worded question because it suggests a link between cell phones and brain cancer. **b)** This question is already assuming the response before it is given. It also gives the advantages of the system without balancing this with any disadvantages. **c)** This question is worded in order to elicit a response in favor of economic incentives for recycling by using the phrases "escalating environmental degradation" and "incipient resource depletion."

3.43 This study would have undercoverage bias as those families just starting out with no children of school age yet would be systematically left out of the study sample.

3.45 Opinion polls can often produce conflicting results when the wording of the statements or questions is subjective. As an example, the phrase "Employees with higher performance must get higher pay" received a 72% agreement in a recent survey on the attitudes Spaniards have toward private business and state intervention. This conflicts with a 71% agreement with the phrase "Everything a society produces should be distributed among its members as equally as possible and there should be no major differences." The first phrase focused on the individual, which appeals to our desire to be rewarded for individual efforts. The second phrase focused on a group (society) and appealed to our desire to have a fair and equitable treatment of society as a whole.

3.47 This is an experiment as the students have a treatment imposed on them. The experimental units are the students. The treatments are the instruction types: paper-based instruction or Web-based instruction. The response variable will be improvement on the standardized drawing test from the pre-test to the post-test. The factor in this experiment is the type of instruction, and the two levels of this factor are paper-based and Web-based. Since this course was in computer graphics technology, there would be a reasonable assumption that the students in the course would be experienced in and interested in the computer based version of drawing. This would not likely generalize to other settings as the same assumptions would not necessarily apply.

3.49 The characteristics of the students who would choose to use the new software might also be characteristics that would lead to higher exam scores like logical thinking abilities, interest in trying new things, and confidence in their ability to learn new things.

3.51

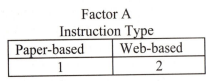

Paper-based	Web-based
1	2

Factor A
Instruction Type

3.53

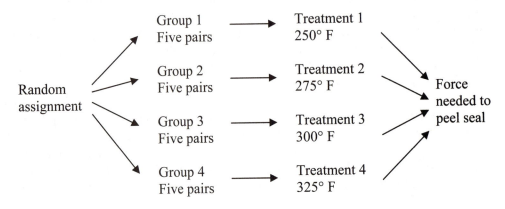

Using Table B at line 140, the 20 pairs could be assigned as follows:

12, 13, 04, 18, 19	Group 1
16, 02, 08, 17, 10	Group 2
05, 09, 06, 01, 20	Group 3
03, 07, 11, 14, 15	Group 4

3.55   In order to ensure that factors influencing electricity use are equal for comparison groups, a control group is needed. Comparing electricity use last year to this year (with the indicator) does not control the factors that may be different between the two years.

3.57   "Significant difference" means the salary differences were so large that it is unlikely they could have been due to chance alone. "No significant difference" means the differences could be due to chance only.

3.59   The experimenter rated their anxiety level both before and after treatment, and the experimenter knew which subjects had received meditation instruction and which had not. Bias could have been introduced if the experimenter had an expectation that meditation would reduce anxiety. It is likely the experimenter unknowingly communicated this expectation to the subjects also.

3.61   Design 1: Randomized Design

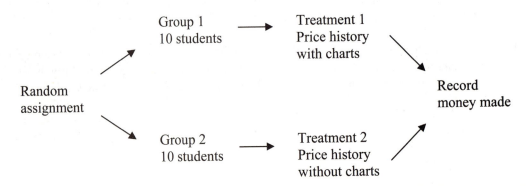

Design 2: Matched Pairs Design
Each student is randomly assigned treatment 1 first or treatment 2 first. The money each student makes using each set of information is compared. The random groups and assignments will vary.

3.63   **a)** Not all subjects have a chance to be in any group. Select subjects using a random number table or computer software to get a random assignment. **b)** Groups of size 4 would not be guaranteed with this method. Instead, use randomization procedures discussed in this chapter to assign subjects to groups. **c)** Rats in different batches may have different characteristics and/or experiences that may affect the results. Randomly select 5 rats from each batch to assign to each treatment option.

3.65   Some blocking variables that might be used are gender, number of years of experience (0–5, 5–10, etc.), and position in company (manager, secretary, etc.). Other answers are also acceptable.

3.67   Comparative experiments that include control groups will best show the effects of the training. Volunteers should be randomly assigned to either the training group or a group receiving no training and just general instructions.

3.69   Applet. Answers will vary.

3.71   **a)** This is a completely randomized design with two treatment groups.

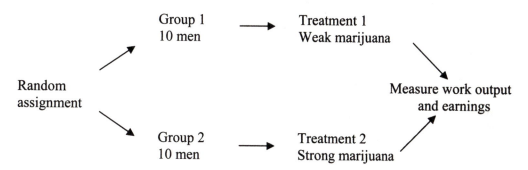

b) Answers will vary using software, but using line 131 on Table B, Group 1 will contain: Dubois (05), Travers (19), Chen (04), Ullman (20), Quinones (16), Thompson (18), Fluharty (07), Lucero (13), Afifi (02), Gerson (08). Group 2 will then contain the remaining subjects: Abate, Brown, Engel, Gutierrez, Hwang, Iselin, Kaplan, McNiell, Morse, and Rosen.

3.73   **a)** Randomly assign your subjects to either Group 1 or Group 2. Each group will taste and rate both the regular and the light mocha drink. However, Group 1 will drink them in the regular/light order, and Group 2 will drink them in the light/regular order. For each group, the taste ratings of the regular and light drinks will be compared, and then the results of the two groups will be compared to see if the order of tasting made a difference to the ratings. To properly blind the subjects, both mocha drinks should be in identical opaque (we are only measuring taste, not appearance) cups with no labels on them.
b) Using line 141 on Table B, the regular/light group will use subjects with labels: 12, 16, 02, 08, 17, 10, 05, 09, 19, and 06. The light/regular group will use the remaining 10 subjects.

3.75 This is an experiment because the students see a treatment (steady price versus price cuts). The explanatory variable is which price history is shown, and the response variable is what price the student expects to pay.

3.77 **a)** The subjects are the 210 children. **b)** The factor is the beverage set and there are three levels. The response variable is the type of drink selected.

c)

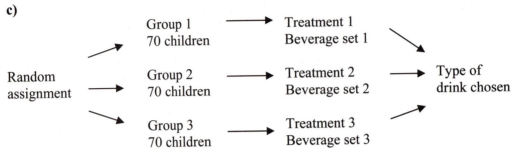

d) Assign the numbers 001 to 210 to the list of children. Starting at line 125 in Table B, the first five subjects for treatment 1 would be: 119, 033, 199, 192, and 148.

3.79 **a)**

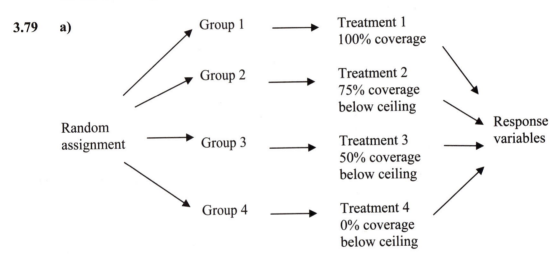

b) Randomly assigning individuals to this type of insurance plan is not an easy task. An individual may not be willing to take part in this experiment if they are assigned to treatments 2, 3, or 4 if these treatments are less than their current coverage. From an ethical perspective, it would be difficult for a company to justify providing less coverage to some individuals than to others.

3.81 **a)**

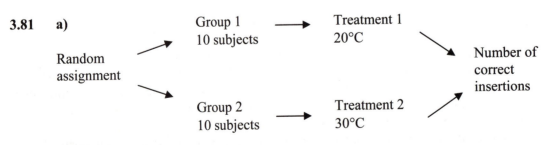

b) A matched pairs experiment would have each subject perform the dexterity test at both temperatures. The order of the treatments would be randomly selected for each subject so that order does not play a role. After each subject performs the dexterity test twice, the difference in the number of correct insertions for each subject would be used to determine if the temperature of the work place has an effect.

3.83 a)

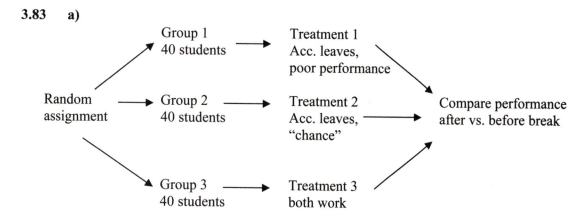

b) Answers will vary using software. Using line 123, the first four subjects for Treatment 1 are: 102, 063, 035, 090.

3.85 This is a voluntary response survey, and, as such, is likely to be biased. Therefore, it would not be accurate to use this to predict population parameters based on this information.

3.87 Both sampling distributions would have the same mean; however, the sampling variability would be greater for the SRS with 200 observations than for the SRS of 400 observations.

3.89 The firm increases its sample size to decrease sampling variability (decrease margin of error).

3.91 a) Population: all U.S. college students. Sample: 17,096 selected students.
b) Population: all restaurant workers. Sample: 100 selected restaurant workers.
c) Population: 584 longleaf pines in the tract. Sample: 40 trees that were measured.

3.93 a) Variability will not change from state to state since variability changes with different sample sizes, not different population sizes. **b)** Using a percent of the state's population results in different sample sizes by state. Changing sample sizes does affect the variability.

3.95 Answers will vary. Increasing the sample size will decrease the spread of the sampling distribution but will not affect the center.

3.97 Applet.

3.99 a) High variability, high bias. **b)** Low variability, low bias. **c)** High variability, low bias. **d)** Low variability, high bias.

3.101 a) Answers will vary. **b)** Be sure the student properly constructs the stemplot or histogram. The center should be close to 0.50.

3.103 Applet.

3.105   **a)** Minimal. **b)** Probably minimal. **c)** Not minimal.

3.107   Answers will vary.

3.109   Answers will vary.

3.111   **a)** Random assignment is not sufficient. There should also be a benefit for the subjects, and the risks to the subjects should be balanced with the benefits. **b)** The board also has the responsibility to follow up on the experiments annually. **c)** Subjects must also give informed consent. Other considerations apply as well such as maintaining the personal rights of the individuals.

3.113   Answers will vary.

3.115   Answers will vary.

3.117   This is not anonymous because it takes place in the person's home. However, it is confidential if the name/address is then separated from the response before the results are publicized.

3.119   **a)** The pollsters must tell the potential respondents what type of questions will be asked and how long it will take to complete the survey. **b)** This is required so the respondents can make sure the polling group is legitimate or so the respondent can issue a complaint if necessary. **c)** Yes, so that readers know whether to question the motivation of the poll, the wording of the question, and the legitimacy of the polling group.

3.121   Psychology 001 uses dependent subjects, which does not seem ethical. The other two courses have acceptable alternatives that make the use of the students more ethical.

3.123   Answers will vary.

3.125   Answers will vary.

3.127   Answers will vary. Students should mention that observational studies do not impose treatments or conditions on the units while experiments do. Experiments often lack reality which is a disadvantage. Observational studies do not allow for control of lurking variables, so results may not be able to be used to demonstrate causation. Other ideas may be included as well.

3.129   NORC pledges to keep all information confidential, including not giving information gathered to anyone else for any purpose. They also state that the information gathered will only be used for statistical purposes.

3.131   Since the environment is manipulated by the researchers and treatments are applied to the students, this is an experiment. The price history seen is the explanatory variable while the price they would expect to pay is the response variable.

3.133   Answers will vary. Check that students have a good idea of the type of study each of these would be beneficial.

3.135 **a)** This is an experiment because the students are reacting to the ads they were shown. The ads are the treatment which is applied to the subjects. **b)** The explanatory variable is the type of ad that is shown to the student, and the response variable is the expected price for the cola that the student states.

3.137 **a)** This is a completely randomized design. It also would be helpful if a standard diet was prescribed to both groups so dietary intake of calcium other than the supplement would be regulated.

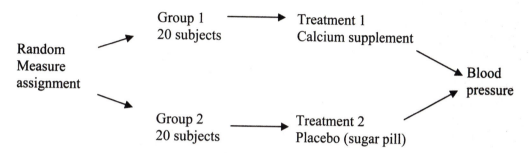

b) Here are the names for the treatment group which will receive the drug, using Table B, starting at line 131 and labeling the subjects in alphabetical order starting with 01.

Chen (05)	Plochman (31)
Rodriguez (32)	Howard (18)
Hruska (19)	Cranston (07)
Bikalis (04)	Fratianna (13)
Liang (25)	Rosen (33)
O'Brian (29)	Asihiro (02)
Imrani (20)	Townsend (36)
Guillen (16)	Krushchev (23)
Tullock (37)	Marsden (27)
Willis (39)	Tompkins (35)

3.139 **a)** Perhaps the job became less stressful coincidentally. Also, maybe the benefit doesn't come just from meditation. Maybe it is from having a routine of doing something positive for mental health on a daily basis. Did the meditation happen during working hours? If so, maybe just having the time away from the work routine was beneficial. **b)** He or she would be looking for improvement in job satisfaction in everyone and might not be objective. Job satisfaction is a very subjective result to measure. **c)** Answers will vary. A proper experimental design might be a completely randomized design. Treatment groups might include the meditation group, a reading group, a walking group, and a group that doesn't do anything special. Another experiment might be a matched pairs design, where pairs of closely matched employees would have one subject do the meditation while the other subject does nothing different for many pairs of employees. The "blind diagnosis" could be if all employees take a standardized questionnaire on paper (without the researcher involved or even in the room) to describe their job satisfaction with only code numbers at the top. The results could then be compiled before matching the code numbers to which the treatment groups

3.141 This is an experiment because there is a treatment imposed: the subjects are asked to taste two muffins and compare the tastes.

3.143   **a)** An experiment to help answer this question would need to compare the number of accidents in an area where most cars have daytime running lights and the number of accidents in an area where most cars do not have daytime running lights. The response variable would be the number of accidents over a specified period of time. **b)** One should be cautious when conducting this type of experiment because, over time, daytime running lights will become less noticeable. It is possible that the number of accidents will drop initially but then increase as the lights become less conspicuous.

3.145   The students can be labeled 0001 to 3478. Starting at line 125, the first six students would be: 1868, 1844, 2351, 1962, 1033, and 3136.

3.147   **a)** The population for this study could be all students enrolled at your school, including part-time students. It makes sense to include part-time students because they purchase college-brand apparel. **b)** A stratified sample makes sense here because there are distinct groups of students enrolled. Gender may be a way to stratify. Full-time or part-time status may be another way to stratify. It might make sense to stratify by whether or not a student is involved in a Greek organization or a member of an athletic team with the reasoning that those students are likely to identify with the school more than others and therefore purchase more college apparel. **c)** Mailing surveys to college students does not ensure a high response rate. Many students move frequently while in college and tend to use their parents' address for mailings. Asking for feedback on this question using the campus newspaper is not a way to collect a random sample. Perhaps a good way to contact a random sample is through the use of email. This may result in a higher response rate.

3.149   "Not significantly different from zero" means that the average returns do not appear to decrease over the first three Mondays in a month. If we compare the average return for the first three Mondays to the average return for the last two Mondays, a "significantly higher average return" means that the last two Mondays combined show a decrease that is not due to chance alone.

3.151   The two factors in this experiment would be time of day and zip code (yes or no). One main post office should be selected, not a drop box. (Drop boxes tend to have only one pick-up time each day.) The time of day factor could have several levels such as 9:00 am, 12:00 pm, and 3:00 pm. There would then be six treatments (3×2). The first treatment would be mailed at 9:00 am with a zip code. There may be variability based on the day of the week the letter is mailed. To ensure equal exposure to this variability, an equal number of letters for each treatment could be mailed on each day of the week.

3.153   Answers will vary.

3.155   The simulations will obviously give various results. Look for the students' understanding of how to use the software to simulate and their understanding of the difference between the numbers of *yes* results (a discrete value) and the $\hat{p}$ values. Note that, for $p = 0.1$, the average should be close to 0.1 and the standard deviation should be close to 0.0387. For $p = 0.3$, the average should be close to 0.3 and the standard deviation should be close to 0.0592. For $p = 0.5$, the average should be close to 0.5 and the standard deviation should be close to 0.0645.

3.157 Subjects are less likely to be truthful in the CAPI survey than in the CASI survey, and therefore the CASI will likely show a higher percent of subjects admitting to drug use.

Case Study 3.1

Answers will vary.

Case Study 3.2

A. A histogram of the entire data set is shown in Figure 1.8. The five-number summary for the entire set of data is 1, 57, 115, 225, 28, 739. This distribution is strongly skewed to the right. There is a large peak close to zero as 7.6% of all calls are 10 seconds long or less. There is also a peak around 50 to 70 seconds.

B. To take a random sample of earnings data from such a large data set you may use the sampling function in Excel under the Tools>Data Analysis toolpak option. Note that Minitab cannot handle a data set this large. The histogram and numerical summaries will be similar to the ones found in part (A).

C. The minimum and maximum are often extreme values within the population. A sample will not necessarily select these precise values. If the maximum and/or the minimum are outliers, we would not expect to even get close to these values with the maximum and minimum of the sample.

Chapter 4: Probability and Sampling Distributions

4.1 Previous experiments estimate that the probability of a head is closer to 0.40 than 0.50. What constitutes a true spin? Do you count spins if the nickel falls off the table? These decisions will affect how you determine this probability.

4.3 Answers will vary. For part (a), six rolls is not a large enough number of trials to give evidence that the die is not fair if you do not get exactly one 6 out of six rolls. The probability of an event is the proportion of times the event occurs in *many* repeated trials of a random phenomenon. For (b) and (c), the probability should be close to 16.67%.

4.5 **a)** 0.105. **b)** Answers will vary but should be close to 0.1.

4.7 The probability should be close to 0.5.

4.9 **a)** 0. **b)** 1. **c)** 0.01. **d)** 0.6.

4.11 Applet.

4.13 **a)** Answers will vary. Most should be between 0.04 and 0.27. The $P(X \geq 14) = 0.1587$. **b)** Histograms will vary but should be fairly symmetrical with center at 0.59 and spread of about 0.66. **c)** The histograms should be fairly symmetrical with a center at 0.65 and a spread of about 0.165. **d)** Both distributions should have similar shapes and the same center, but the spread of the distribution of 320 observations should be much smaller.

4.15 **a)** Answers will vary but should be close to 6%. **b)** This simulation represents 100 randomly selected items which have been purchased. A success is an item which is returned. Over the long run, there is a 0.06 probability that an item which is purchased will be returned.

4.17 **a)** There is some variation here depending on the unit of measure. If we measure in whole hours $S = \{0, 1, ..., 24\}$. **b)** $S = \{0, 1, 2, ..., 11,000\}$. **c)** $S = \{0, 1, 2, ..., 12\}$. **d)** $S = (0, \infty)$. **e)** $S = \{\text{integers in the interval } [0, \infty)\}$.

4.19 **a)** 0.177. **b)** 0.823.

4.21 Model 1 is not legitimate because the probabilities do not add up to 1. Model 2 is legitimate. Model 3 is not legitimate because the probabilities add up to more than 1. Model 4 is not legitimate because the probabilities are all greater than or equal to 1.

4.23 **a)** 0.48 because the probabilities must add up to 1. **b)** 0.09.

4.25 **a)** Area $= \dfrac{1}{2}bh = \dfrac{1}{2}(2)(1) = 1$.

b) $P(T < 1) = 0.5$

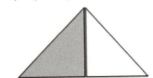

c) $P(T < 0.5) = 0.125$.

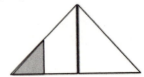

4.27   **a)** $P(Y > 515) = 0.50$.   **b)** $P(Y > 631) = 0.16$.

4.29   **a)** The probabilities are 0.09 for small businesses and 0.37 for big businesses.
b) The probabilities are 0.91 for small businesses and 0.63 for big businesses.

4.31   **a)** Sum of probabilities is 1.00. All probabilities are between 0 and 1.   **b)** 0.89.

4.33   **a)** The probabilities are 0.8 for Canada and 0.902 for the United States.
b) The probabilities are 0.187 for Canada and 0.082 for the United States.

4.35   **a)** $P(Y > 1) = 0.69$.   **b)** 0.28.   **c)** 0.65.

4.37   **a)** 0.08.   **b)** 0.93.   **c)** 0.07.

4.39   **a)** 0.12.   **b)** 0.5.

4.41   **a)** Legitimate.   **b)** Not legitimate, sum > 1.   **c)** Not legitimate, sum < 1.

4.43   **a)** $P(A) = 0.29$, $P(B) = 0.18$.   **b)** "The farm is at least 50 acres," $P(A^c) = 1 - P(A) = 0.71$.
c) "The farm is less than 50 acres or at least 500 acres," $P(A \text{ or } B) = 0.47$.

4.45   **a)** NNN, NNO, NON, ONN, NOO, ONO, OON, OOO. 1/8.   **b)** 3/8.   **c)** $X = \{0, 1, 2, 3\}$,
$P(0) = 1/8$, $P(1) = 3/8$, $P(2) = 3/8$, $P(3) = 1/8$.

4.47   **a)** Verify that probabilities add to 1. All probabilities are between 0 and 1.
b) 0.11.   **c)** 0.04.   **d)** 0.32.   **e)** 0.75.   **f)** $P(X > 2) = 0.43$.

4.49   **a)** 0.3821.   **b)** 0.1768.

4.51   **a)** Continuous.   **b)** Discrete.   **c)** Continuous.   **d)** Discrete.

4.53   **a)** Verify that probabilities add to 1. All probabilities are between 0 and 1.

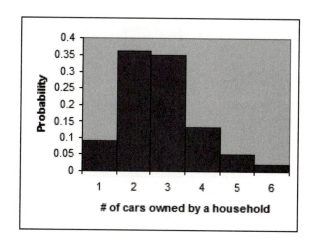

b) $P(X \geq 1)$ means "What is the probability that a household owns at least one car?" This
equals 0.91.   **c)** 0.20.

4.55   **a)** 0.50.   **b)** 0.0344.   **c)** 0.0344.

4.57   The mean, μ, of hard-drive size for laptop computers is 169.5 GB. This is the expected size of hard drives for purchases over a long period of time. This is not useful because hard drives don't actually come in this size.

4.59   **a)** $\mu_X = 1.9$. $\mu_Y = 3.05$. **b)** $\mu_{8000X} = \$15,200$. $\mu_{30000Y} = \$91,500$. **c)** $\mu_{X+Y} = 4.95$. $\mu_{8000X + 30000Y} = \$106,700$.

4.61 $\sigma^2_Y = 19225$, $\sigma_Y = 138.65$.

4.63   **a)** $\sigma^2_X = 0.49$, $\sigma_X = 0.7$. **b)** $\sigma^2_Y = 1.048$, $\sigma_Y = 1.023$.

4.65 $\mu_{X-Y} = \$100$; $\sigma_{X-Y} = \$100$. When two variables have a positive correlation, if one variable increases, the other will also increase. If one variable decreases, the other will also decrease. This results in the average differences being smaller than if they were independent. With independence, if one variable increases, the other could increase or decrease.

4.67   **a)** 720. **b)**

No. of Transactions	0	1	2	3	4	5
Percent of Clients	0.08	0.17	0.25	0.19	0.18	0.13

c) Yes, if for 5 transactions the 0.125 is rounded to 0.13, then the probabilities add up to 1. If the 0.125 is rounded down to 0.12, the probabilities add up to 0.99.

4.69   **a)** 0.0010. **b)** 0.1131. **c)** 394.24 bets per second.

4.71   **a)** –1280 dollars. **b)**

X	$\$218,720$	$-\$1,280$
P(X)	0.003	0.997

c) A histogram will show two bars, one with height very close to zero and one with height very close to 1.

4.73   **a)**

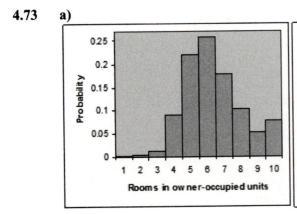
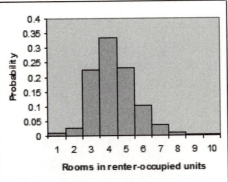

The distribution of number of rooms in owner-occupied units has more mass toward the higher end, 6 through 10 rooms. It appears that the center of the distribution is at 6. The distribution on number of rooms in renter-occupied units is slightly skewed right with a central tendency of 4.

b) $\mu_o = 6.39$, $\mu_r = 4.39$.

4.75 The distribution on number of rooms in owner-occupied units appears to be more spread out than number of rooms in renter-occupied units. $\sigma_o = 1.69$, $\sigma_r = 1.42$.

4.77 Answers will vary as data sets are randomly generated. The average should be close to 5000.

4.79 Let X = expected payoff for a $1 bet on the box. X = $83.33 or $0. $P(\$83.33) = 0.006$, $P(\$0) = 0.994$. $\mu_X = \$0.50$.

4.81 **a)** Independent. **b)** Independent. **c)** Dependent.

4.83 **a)** Let X = time to bring part from bin to chassis and Y = time to attach part to chassis. $\mu_{X+Y} = 31$ seconds. **b)** No. **c)** There would be no change.

4.85 $\sigma_{X+Y} = 4.47$ seconds. If dependent, $\sigma_{X+Y} = 4.98$ seconds. Positive correlation means that, when one variable increases, the other also increases. The variation then builds to result in increased variation on their sum.

4.87

X	$P(X)$
$\mu - \sigma$	0.5
$\mu + \sigma$	0.5

$$\mu_X = 0.5(\mu - \sigma) + 0.5(\mu + \sigma) = \mu, \quad \sigma^2_X = 0.5(\mu - \sigma)^2 + 0.5(\mu + \sigma)^2 - \mu^2 = \sigma^2$$

4.89 **a)** $\sigma_{X+Y} = \$2796$. **b)** $\sigma_{2000X + 3000Y} = \$5,606,739$.

4.91 **a)** The students' scores should not be influenced by each other. **b)** $\mu_{F-M} = 15$, $\sigma_{F-M} = 44.82$. **c)** No, you cannot calculate a precise probability because, even though we know the mean and standard deviation, we do not know the shape of the probability distribution.

4.93 **a)** $\mu_{Y-X} = 0.87$, $\sigma_{Y-X} = 11.77$. **b)** $P(Y - X > 0) = 0.5279$.

4.95 With $\rho = 0$, $\sigma_{WY} = 3.35\%$. There would be no change in the mean.

4.97 For $\rho = 1.0$, $\sigma^2_{X+Y} = \sigma^2_X + \sigma^2_Y + 2\sigma_X\sigma_Y = (\sigma_X + \sigma_Y)^2$; therefore, $\sigma_{X+Y} = \sigma_X + \sigma_Y$.

4.99 **a)** $\mu_X = 550°$ and $\sigma_X = 5.7°$. **b)** $\mu_{X-550} = 0°$ and $\sigma_{X-550} = 5.7°$. **c)** $\mu_Y = 1022°F$ and $\sigma_Y = 10.26°F$.

4.101 Parameters describe fixed populations. If we treat the set of tasks completed for this study as a sample of all tasks that could be performed on a computer, then these numbers represent statistics.

4.103 Each time Joe plays, his return is either 0 or 600. If he averages his winnings over many years, he'll find an average very close to $0.60.

4.105 The "law of averages" assumes independent trials. This means that each time Tony is at bat, his chance of getting a hit is the same. Therefore, Tony is not "due" a hit.

4.107 **a)** $\mu = 69.4$. **b)** Answers will vary. One sample is 7, 3, 6, 4 with an $\overline{x} = 65.25$.
c) The histogram below is one example.

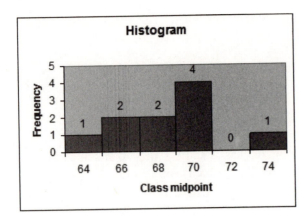

4.109 **a)** $\overline{x}$ is an unbiased estimator of μ because, if we were to take many, many samples and calculate many, many values of $\overline{x}$, the average of all the $\overline{x}$'s would be equal to μ, the mean of the population. **b)** The law of large numbers tells us that, when we take larger and larger samples, the value of $\overline{x}$, the sample mean, gets closer and closer to μ, the population mean.

4.111 Approximately 0.0.

4.113 500,000,000 is a parameter, and 5.6 is a statistic.

4.115 The number 19 is a parameter and 14 is a statistic.

4.117 **a)** $\overline{x}$ has an approximate Normal distribution with a mean of 123 mg and a standard deviation of 0.0462. **b)** $P(\overline{x} > 124) \approx 0$.

4.119 $P(\overline{x} > 210.53) = 0.0052$.

4.121 0.0125.

4.123 Between 0.1108 and 0.1892 average defects.

4.125 **a)** $P(\overline{x} > 400) \approx 0$. **b)** We need the shape of the population distribution. The central limit theorem allowed us to approximate the probability in part (a).

4.127 **a)** $E(X) = -\$620$ and $\sigma = \$12,031.81$. **b)** $E(Y) = \$219,340$ and $\sigma = \$12,031.81$. **c)** The standard deviation is the same for both.

4.129 **a)** 0.0866. **b)** 9. **c)** There is always sampling variability, but this variability is reduced when an average is used instead of an individual measurement. The larger the sample size, the smaller the variability.

4.131 **a)** 0.3544. **b)** 0.8558. **c)** 0.9990. **d)** The probabilities using the central limit theorem are more accurate as the sample size increases. The 150-bag probability calculation is probably fairly accurate, but the 3-bag probability calculation is probably not.

4.133 0.55.

4.135 **a)** 0.1. **b)** No, there is also a probability of 0.10 that a person either has no opinion or thinks that this is debatable.

4.137 **a)** Verify that probabilities add to 1 and that all probabilities are between 0 and 1. **b)** 0.43. **c)** 0.96. **d)** 0.28. **e)** 0.72.

4.139

Y	$P(Y)$
1	3/36
2	3/36
3	3/36
4	3/36
5	3/36
6	3/36
7	3/36
8	3/36
9	3/36
10	3/36
11	3/36
12	3/36

4.141 Let Y be the weight of the carton. $Y \sim N(780, 17.32)$. $P(750 < Y < 825) = 0.9535$.

4.143 **a)** Verify that sum of probabilities is 1 and all probabilities are between 0 and 1. **b)** $\mu = 2.45$.

4.145 A correlation of 0 would result in the smallest standard deviation.

4.147 90% of the time the mean monthly return will be greater than -0.0411 for Dell, -0.0383 for Apple and -0.0306 for the portfolio.

4.149 If you sold only 10 policies, you would probably lose a lot of money. Even though the average loss per person is only $400, that average comes from a distribution that has some losses equal to zero and a very few losses that are greater, perhaps $200,000 or more. One loss would be much more than the profit from nine policies. If you sold thousands of policies, the gain from the many, many policies that paid out $0 would be much more than the few losses incurred.

4.151 $P(\$1250) = 0.99058$, $\mu_x = \$303.35$.

4.153 $\sigma_x = \$9707.6$.

4.155 **a)** $S = \{1, 2, 3, ..., 50\}$. **b)** Discrete because the variable takes on integer values. **c)** There are 50 possible values.

4.157 **a)** $(0, 35]$. **b)** Continuous because the variable is described on intervals. **c)** Infinite.

Chapter 5: Probability Theory

5.1 **a)** It is reasonable to assume high school ranks are independent because a student's performance is not influenced by another student's high school performance.
b) $(0.41)(0.41) = 0.1681$. **c)** $(0.41)(0.01) = 0.0041$.

5.3 **a)** 0.421. **b)** 0.579.

5.5 $P(\text{lights work for 3 years}) = P(\text{no lights fail in 3 years}) = (1 - 0.02)^{20} = 0.6676$.

5.7 **a)** 15%. **b)** 20%.

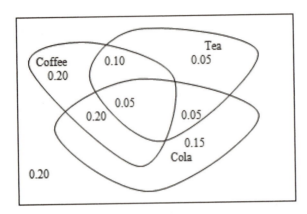

5.9 **a)** $W = \{0, 1, 2, 3\}$.

W	Arrangements	Part (b) Probability of Each Arrangement	Part (c) Probability for W
$W = 0$	DDD	$(0.73)^3 = 0.389$	0.389
$W = 1$	DDF	$(0.27)(0.73)^2 = 0.144$	0.432
	DFD	$(0.27)(0.73)^2 = 0.144$	
	FDD	$(0.27)(0.73)^2 = 0.144$	
$W = 2$	DFF	$(0.27)^2(0.73) = 0.0532$	0.160
	FDF	$(0.27)^2(0.73) = 0.0532$	
	FFD	$(0.27)^2(0.73) = 0.0532$	
$W = 3$	FFF	$(0.27)^3 = 0.0197$	0.197

5.11 **a)** $(0.09)^6 = 0.0000005$. **b)** $(0.91)^6 = 0.568$. **c)** $6(0.91)^5(0.09) = 0.337$.

5.13 $P(\text{win at least once}) = 1 - P(\text{lose all five times}) = 1 - (0.98)^5 = 0.0961$.

5.15 **a)** $(0.65)^3 = 0.2746$. **b)** Since we assume the years are independent, the third year has probability of 0.65 of going up. **c)** $P(\text{up two years in a row or down two years in a row}) = (0.65)(0.65) + (0.35)(0.35) = 0.545$.

5.17 $P(A \text{ or } B) = 0.6 + 0.5 - 0.3 = 0.8$.

5.19   If $P(A)P(B) = P(A$ and $B)$ then A and B are independent. We were given $P(A)$, $P(B)$, and $P(A$ and $B)$. $P(A)P(B) = (0.6)(0.5) = 0.3$. This is equal to $P(A$ and $B)$; therefore, A and B are independent of each other.

5.21   No. $P(A$ or $B) = P(A) + P(B) - P(A$ and $B)$, and $P(A$ and $B)$ will not equal zero since both $P(A)$ and $P(B)$ are both greater than zero.

5.23   **a)** $P($throwing an 11$) = 2/36 = 0.05556$, $P($throwing three 11s in a row$) = (0.05556)^3 = 0.000172$. **b)** Using the probability values from part (a) and the formula given, the odds against throwing an 11 are 17 to 1. The odds against throwing three 11s in a row are 5812 to 1. The writer is correct about his first statement, but not about his second statement. He should have multiplied $18 \times 18 \times 18$. Note that if we use 1/18 as the probability of rolling an 11, then the odds of throwing three 11s are calculated exactly to be $18^3 - 1$, or 5831 to 1. In only one out of 5832 tries will we throw three 11s in a row.

5.25 $P(A$ and $B) = 0.1472$.

5.27   **a)** 0.6946. **b)** 0.7627. **c)** 0.2921. **d)** 0.0158. **e)** 0.1884.

5.29   **a)** $P(B \mid A) = 0.5833$. **b)** $P(A \mid B) = 0.3621$. **c)** No, $P(A)P(B) \neq P(A$ and $B)$.

5.31 $P(B$ and $A) = (0.04)(0.45) = 0.018$. $P(B$ and $A^c) = (0.01)(0.55) = 0.0055$. $P(B) = 0.018+0.0055$. $P(A \mid B) = 0.766$.

5.33 $P($premium $\mid$ pay at least \$30$) = 0.337$.

5.35 $P(O \mid D) = 0.8989$. Given that the customer defaults on the loan, there is an 89.89% chance that the customer overdraws the account. If there is a 25% chance that the customer will overdraw, then $P(O|D) = 0.8163$.

5.37   **a)** $P(G \mid C) = 0.25$. **b)** $P(G \mid C^c) = 0.3333$.

5.39   0.0043.

5.41

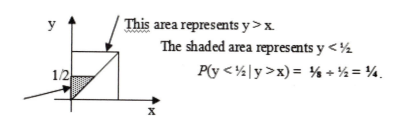

5.43

	Did Not Finish High School	High School	Some College	Bachelor's or Higher
Unemployment Rate	0.0714	0.0436	0.0355	0.0202

As level of education increases, unemployment rate decreases. Since the probability of being unemployed changes with different given education levels, the variables are not independent.

5.45

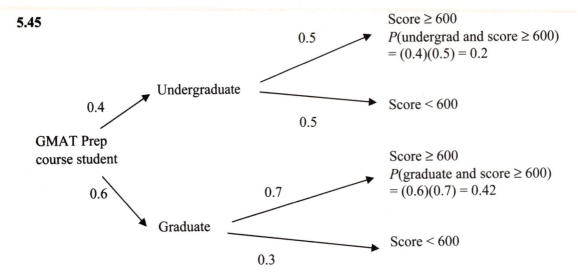

a) P(undergrad and score $\geq$ 600) = 0.2, P(graduate and score $\geq$ 600) = 0.42.

b) P(score $\geq$ 600) = 0.62.

5.47 P(undergraduate | score $\geq$ 600) = 0.3226.

5.49   Let D = event that a credit card customer defaults and L = event that a credit card customer is late for two or more monthly payments. **a)** $P(D \mid L) = 0.0637$. **b)** Between 93% and 94% of customers who have their credit denied will *not* default on their payments. **c)** No. Only 3% of the customers default. Of those who are late, only 6.37% default. Knowing that a customer is late on payments does not dramatically increase the chance that they will default on their payments.

5.51   Yes, $P(A)P(B) = P(A \text{ and } B)$.

5.53   Let NC = event that an item is nonconforming and C = event that an item is conforming. Let I = event that an item was completely inspected. $P(NC) = 0.08$, $P(C) = 0.92$, $P(I \mid NC) = 0.55$, $P(I \mid C) = 0.20$. $P(NC \mid I) = 0.1930$.

5.55 X does not have a Binomial distribution. There is not a fixed number of trials.

5.57   **a)** $X = \{0, 1, 2, 3, 4, 5\}$.

b)

X	0	1	2	3	4	5
$P(X)$	0.4207	0.3977	0.1504	0.0284	0.0027	0.0001

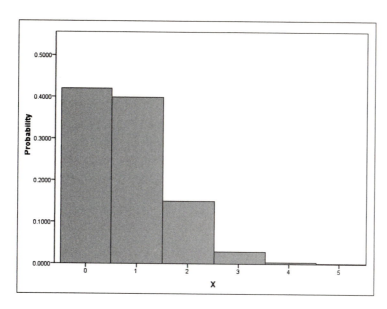

5.59   0.0074.

5.61 $\mu = 8$, $\sigma = 2.19$.

5.63   **a)** $\mu = 16$. **b)** $\sigma = 1.79$. **c)** $\sigma = 1.34$. $\sigma = 0.445$. As p gets closer to one, the standard deviation gets smaller.

5.65   **a)** Using the Normal approximation to the Binomial, $P(X \geq 100) = P(Z \geq 2.89) = 0.0019$. **b)** Sample size can greatly influence the probabilities, which in turn can influence your conclusions. The larger the sample size, the more noticeable the differences between two populations.

5.67   **a)** $\mu_x = np = 6$, $\mu_{\hat{p}} = p = 0.5$. **b)** $\mu_x = 60$ if $n = 120$, and $\mu_x = 600$ if $n = 1200$. $\mu_{\hat{p}}$ stays the same regardless of sample size.

5.69   **a)** X is Binomial with $n = 1555$ and $p = 0.20$. **b)** Using the Normal approximation to the Binomial for proportions, $\mu_X = 311$, $\sigma_X = 15.77$, and $P(X < 301) = P(Z < -0.63) = 0.2643$. Using software for the Binomial distribution, $P(X \leq 300) = 0.254$.

5.71   **a)** X does not have a Binomial distribution. The probability of performing satisfactorily on the exam will be different for each machinist. **b)** X does have a Binomial distribution. There is a fixed number of trials: 100. Each trial is independent of the others. There are only two possible outcomes for each trial: yes or no. If we assume there is a fixed proportion of people in the population who choose to take part in studies, then the probability the individual will say yes is the same for each person.

5.73   **a)** $n = 10$, $p = 0.25$ **b)** $P(X = 2) = 0.2816$. **c)** $P(X \leq 2) = 0.5256$. **d)** $\mu = 2.5$, $\sigma = 1.37$.

5.75   **a)** $n = 5$, $p = 0.65$. **b)** 0, 1, 2, 3, 4, or 5. **c)** See the graph below. **d)** $\mu = 3.25$, $\sigma = 1.07$.

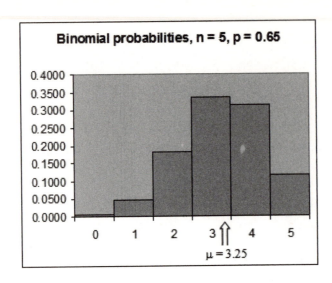

5.77 **a)** $\mu = 75$, $\sigma = 4.33$, $P(X \le 70) = 0.1251$. **b)** A score of 70% on a 250 question test is 175 correct answers. $\mu = 187.5$, $\sigma = 6.85$, $P(X \le 175) = 0.0344$.

5.79 **a)** It is reasonable to use the Binomial distribution to the number who respond because you have a fixed number of trials, they can be assumed to be independent, there are only two possible outcomes (either they respond or they do not) and the probability of a response stays the same for each trial. **b)** $\mu = 75$. **c)** $P(X \le 70) = 0.2061$. **d)** $n = 200$.

5.81 **a)** $\mu = 180$, $\sigma = 12.6$. **b)** $P(X \le 170) = 0.2148$.

5.83 With the continuity correction, the probability is slightly larger at 0.0069. This value is closer to the exact probability than the original approximation.

5.85 **a)** Poisson with $\mu = 84$. **b)** $P(X \le 66) = 0.0248$.

5.87 **a)** The employees are independent and each equally likely to be hospitalized. **b)** 0.0620. **c)** 0.938. **d)** 0.256.

5.89 **a)** 0.0821. **b)** 0.2424.

5.91 **a)** $P(X \ge 50) = 1 - P(X \le 49) = 0.445$. **b)** $\sigma = 6.98$, $\sigma = 9.87$. **c)** Using $\mu = 97.4$, $P(X \ge 100) = 1 - P(X \le 99) = 0.4095$.

5.93 **a)** $\sigma = 3.873$. **b)** $P(X \le 10) = 0.1185$. **c)** $P(X > 30) = 1 - P(X \le 30) = 0.0002$.

5.95 **a)** $P(X \ge 5) = 0.9982$. **b)** $P(X \ge 5) = 0.827$. **c)** 0.2746.

5.97 **a)** $\sigma = 1.52$. **b)** $P(X > 5) = 1 - P(X \le 5) = 0.03$. **c)** $P(X > 3) = 0.20$ and $P(X > 4) = 0.0838$, so $k = 3$.

5.99 (14,632.58, 15,367.42).

5.101 **a)** 0.1416. **b)** 0.1029. **c)** The people are independent from each other, and each person has a 70% chance of being male.

5.103   0.75.

5.105   Use $P(I) = P(I \text{ and } F) + P(I \text{ and } S)$ and $P(S) = P(I \text{ and } S) + P(I^c \text{ and } S)$. $P(I^c \text{ and } S) = 0.84$.

5.107   **a)** 0.5649. **b)** 0.2377. **c)** 0.0483. **d)** 0.6677.

5.109   **a)**

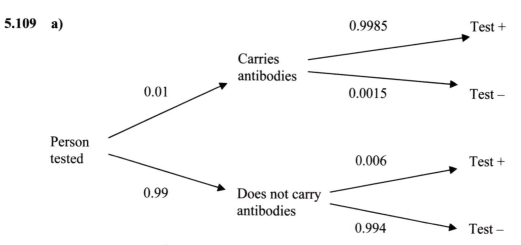

b) $P(\text{test} +) = 0.01(0.9985) + 0.99(0.006) = 0.0159$.
c) $P(\text{carry antibodies} \mid \text{test} +) = 0.627$.

5.111   **a)** 0.0099. **b)** 0.0098. **c)** 0.0097. 0.0096. $P(\text{first defective product is the } k\text{th one}) = p(1-p)^{k-1}$.

5.113   **a)** 0.799. **b)** 0.921.

5.115   **a)** 0.151. **b)** $P(H \mid B) = 0.0465$.

5.117 $P(\text{iMac}) = 0.05$, $P(\text{did not buy an iMac}) = 0.95$, $P(\text{First time} \mid \text{iMac}) = 0.32$, $P(\text{First time} \mid \text{did not buy iMac}) = 0.40$, $P(\text{iMac} \mid \text{First time}) = 0.04$. Approximately 4% of first time computer buyers bought an iMac.

5.119   **a)** $\mu = 3.75$. **b)** 0.000795. **c)** 0.034.

5.121   **a)** $\mu = 1250$. **b)** 0.5596.

5.123   **a)** It is reasonable to use the Binomial distribution because there are a fixed number of trials (n), the trials can be considered independent, there are only two possible outcomes, and the probability stays the same for each trial (since we are observing at the same location on the same day). **b)** It is more likely that the male will be driving after a dance on campus than after church on Sunday. **c)** 0.4557. **d)** 0.1065.

5.125   No, the Poisson distribution is not appropriate here since the probability of a success is not the same for all equal size intervals.

5.127 $P(X \geq 1) = 1 - P(X = 0) = 0.0334$.

5.129 $P(X \geq 19) = 1 - P(X \leq 18) = 1 - 0.8195 = 0.1805$.

Case Study 5.1

Verify the calculations. **a)** P(one or more errors in 365 days) =
$1 - (1 - (1/9000000000))^{(365 \times 1000)} = 0.000041$.
b) P(one or more errors in 365 days) $= 1 - (1 - (1/100000000))^{(365 \times 4200000)} = 0.9999998$.

Case Study 5.2

a) P(one or more errors in 24 days) $= 1 - (1 - (1/100000000))^{(24 \times 4200000)} = 0.6351$.
Based on this probability it is likely that an individual will see an error in 24 days. There is a better than 50% chance of seeing an error. **b)** P(one or more errors in one day) = 0.04113. For 100,000 users the expectation is over 4000 mistakes.

Chapter 6: Introduction to Inference

6.1 $\sigma_{\bar{x}} = \$22.$

6.3 $44.

6.5 ($394.70, $415.34).

6.7 $n = 1244.68$, so use a sample size of 1245.

6.9 **a)** The manager did not divide the standard deviation by the square root of the sample size. **b)** The confidence interval is used to estimate the location of the population mean, not the location of the sample mean. **c)** The confidence interval only estimates the location of the population mean, not the location of the overall population. **d)** The distribution of the sample mean will be approximately Normal, not the distribution of the population.

6.11 As the sample size increases, the width of the confidence interval decreases. The center of the confidence interval remains unchanged.

n	10	20	40	100
$\bar{x} \pm m$	50 ± 3.10	50 ± 2.19	50 ± 1.55	50 ± 0.98
CI	(46.90, 53.10)	(47.81, 52.19)	(48.45, 51.55)	(49.02, 50.98)

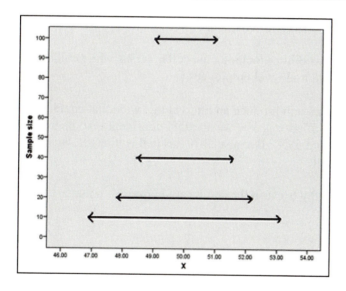

6.13 The students in your major will have a smaller standard deviation because many of them will be taking the same classes which require the same textbooks. The smaller standard deviation leads to a smaller margin of error.

6.15 **a)** A 100% confidence interval would contain every potentially possible value for the population mean. This interval would be so wide that it would not provide any information about the center of the distribution at all. **b)** A 0% confidence interval would just include the sample mean. The probability that the population mean is exactly equal to the sample mean is so small that we would expect this to happen 0% of the time.

6.17 $\bar{x} = 343.165$, the average of the upper and lower bounds of the confidence interval. The margin of error for the 90% confidence interval is 13.295. The 95% confidence interval would have a margin of error of 15.841, resulting in a confidence interval of (327.32, 359.01).

6.19 When there is non-response bias, the accuracy will be in question, regardless of the size of the margin of error.

6.21 9.29 ± 0.92 or (8.37, 10.21).

6.23 **a)** (101.75, 128.25). **b)** No, this is an interval that describes the average, not a single value.

6.25 **a)** 136.099 pounds. **b)** 2.02 pounds. **c)** (132.137, 140.061).

6.27 (11.03, 12.57).

6.29 $n = 74.37$, which we would round up to a sample size of 75 shoppers.

6.31 **a)** No. We are 95% confident that the interval contains the true population percent. **b)** This particular interval was produced by a method that will give an interval that contains the true percent of the population that like their job 95% of the time. When we apply the method once, we do not know if our interval correctly includes the population percent or not. Because the method yields a correct result 95% of the time, we say we are 95% confident that this is one of the intervals that correctly includes the population percent. **c)** 1.531. **d)** No, the margin of error only covers random variation.

6.33 Answers will vary. Some possibilities include blue collar versus white collar jobs, management versus entry- and mid-level employees.

6.35 **a)** 95 out of 100 of the samples will produce an interval this wide that contains the true percentage. **b)** If the margin of error is 3%, we are 95% confident that the true percentage is between 49% and 55%. Because 50% lies in this interval, we conclude that the election is too close to call.

6.37 No, this result is not trustworthy because the results are biased. A voluntary response sample was used.

6.39 $H_0: \mu = 0$; $H_a: \mu < 0$.

6.41 **a)** $z = -1.58$.

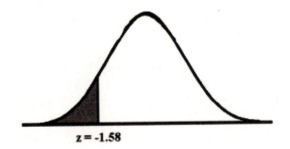

$z = -1.58$

b) *P*-value = 2 × 0.0571 = 0.1142. No, this is not strong evidence that Cleveland is different than the national average.

6.43   You must state what you want the evidence to support *before* you take the sample, not after sampling.

6.45 $|z| \geq 2.81$.

6.47 $\alpha = 0.0456$ using Table A or $\alpha = 0.05$ using the 68-95-99.7 rule. $\alpha = .0026$ using Table A or $\alpha = .003$ using the 68-95-99.7 rule.

6.49   **a)** 0.0287. **b)** 0.9713. **c)** 0.0574.

6.51   **a)** Yes, a 95% confidence interval is the same as conducting a two-sided test using an $\alpha = 0.05$. Since the *P*-value is greater than 0.05, the results of the test are not significant and we can conclude that 10 would be in the confidence interval. **b)** No, using the same reasoning as above. A 90% confidence interval is the same as conducting a two-sided test using an $\alpha = 0.10$. Since the *P*-value is less than 0.10, the results of this test are significant and we can conclude that 10 would not be in the confidence interval.

6.53   **a)** Yes. **b)** No. **c)** The *P*-value is greater than 0.01 but less than 0.05. If the *P*-value is ≤ the significance level, then the null hypothesis can be rejected.

6.55   **a)** The null hypothesis should be $\mu = 0$. The alternative hypothesis could be $\mu > 0$. The null hypothesis always has the equal sign and represents "no effect." **b)** The standard deviation of the sample mean should be $18/\sqrt{30}$. The square root was left out. **c)** The hypothesis test uses the population parameter μ, not the sample statistic $\bar{x}$. Hypothesis tests and confidence intervals always give us information about the population parameters, not the sample statistics.

6.57   If a significance level of 0.05 is used, do not reject the null hypothesis. There is not enough evidence to say that students in the West region have significantly different credit card debt than from those in the South region.

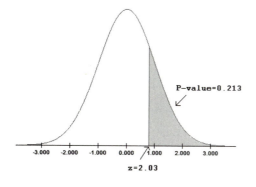

6.59   **a)** $H_0\colon \mu = 31$; $H_a\colon \mu > 31$. **b)** $H_0\colon \mu = 4$; $H_a\colon \mu \neq 4$. **c)** $H_0\colon \mu = 1400$; $H_a\colon \mu < 1400$.

6.61   **a)** $H_0\colon p_M = p_F$; $H_a\colon p_M > p_F$, where the parameters of interest are the percent of males, p_M, and the percent of females, p_F, in the population who name economics as their favorite subject. **b)** H_0: $\mu_A = \mu_B$; H_a: $\mu_A > \mu_B$, where μ_A is the mean score on the test of basketball skills for the population of all sixth-grade students if all were treated as those in group A and μ_B is the mean score on the test of basketball skills for the population of all sixth-grade students if all were treated as those in group B. **c)** H_0: $\rho_A = 0$; H_a: $\rho > 0$, where the parameter of interest is the correlation ρ between income and the percent of disposable income that is saved by employed young adults.

6.63   Customers who had complaints handled satisfactorily were no more or less likely to purchase a new vehicle of the same make than those who did not complain. Any difference between these two percents was small enough that it could be just due to the selection of people who were chosen. However, customers who had complaints handled unsatisfactorily were significantly less likely to repurchase the same make of vehicle than those who did not complain or who complained but received good resolution. The *P*-value tells us that the chance that these percents would be this different just due to random chance is less than 1 out of 1000. With a *P*-value lower than 0.001, this indicates that we can be extremely confident that there is a significantly lower percent of those with a negative resolution to their complaint who plan to purchase the same make again.

6.65   **a)** H_0: $\mu_A = \mu_B$ and H_a: $\mu_A \neq \mu_B$, where group A are students who exercise regularly and group B are students who do not exercise regularly. **b)** No, the *P*-value is large; therefore, this sample result does not give evidence in favor of the alternative hypothesis. **c)** It would be good to know how the sample was collected and if this was an observational study or a designed experiment.

6.67 $z = 3.56$. *P*-value ≈ 0. The sensible conclusion is that the poems were written by a different author.

6.69 $z = 2.40$. *P*-value $= 0.0164$. Yes, since the *P*-value is so small, this sample provides very strong evidence against H_0. There is evidence that the population mean corn yield is not 135 bushels per acre. Since $n = 50$, this sample size is large enough to overcome a slightly non-Normal population.

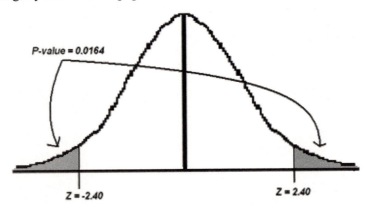

6.71   **a)** No, *P*-value $= 0.2040$, which is greater than 0.05. **b)** No.

6.73   *P*-value < 0.001.

6.75   **a)** $z \geq 1.645$. **b)** $|z| \geq 1.96$. **c)** Part (a) is a one-sided test and part (b) is a two-sided test.

6.77 **a)** (99.038, 109.222). **b)** H_0: $\mu = 105$; H_a: $\mu \neq 105$. Since we are 95% confident that μ is between 99.038 and 109.222, we cannot conclude that the null hypothesis is false.

6.79 **a)** H_0: $\mu = 7$; H_a: $\mu \neq 7$. Yes, the interval does not contain 7; therefore, it is reasonable to conclude that the null hypothesis is false. **b)** No, 5 is in the interval.

6.81 **a)** No, the *z* statistic is 1.64, which is less than the critical value of 1.645. **b)** Yes, the *z* statistic is 1.65, which is greater than the critical value of 1.645.

6.83 Answers will vary. Suppose we are testing a new soft drink in a supermarket. People will be influenced by comments from other shoppers tasting the soda. Results from this type of sample are probably biased.

6.85 Statistical significance and practical significance are not necessarily the same thing. With a significance level this high, we would be willing to reject the null hypothesis even when no practical effect exists. A *P*-value of 0.5 corresponds to a *z* value of 0, which means that our sample statistic would be 0 standard deviations away from the null hypothesis mean, but we would still reject the null hypothesis.

6.87 Answers will vary.

6.89 **a)** No. **b)** Yes. **c)** No.

6.91 **a)** *P*-value = 0.3821. **b)** *P*-value = 0.1711. **c)** *P*-value = 0.0013.

6.93 No, the sample was a voluntary response sample, which can lead to biased results.

6.95 The *P*-values 0.008 and 0.001 are statistically significant.

6.97 **a)** The distribution of *X* is Binomial. **b)** $P(X \geq 2) = 0.9027$.

6.99 As the sample size increases, the power increases, assuming everything else about the studies is the same.

6.101 Since 80 is farther away from 50 than 70 is, the power will be higher than 0.5.

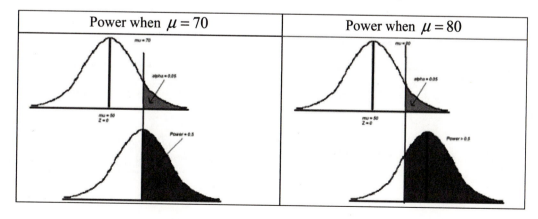

6.103 **a)** A sample size of at least 35 is required. **b)** This curve only shows positive differences since we are looking at a one-sided alternative hypothesis.

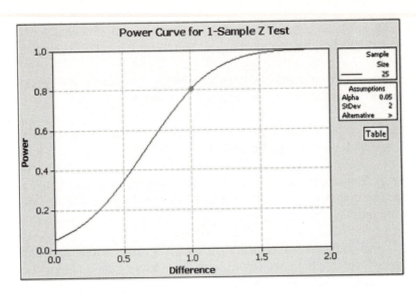

c) Differences over about 1.5 result in almost certain detection of the alternative.

6.105 1. Increase the significance level. 2. Increase the sample size. 3. Decrease the population standard deviation. 4. Consider an alternative that is farther away from μ_0.

6.107 **a)** $0.25n$. **b)** Smaller sample sizes mean increased variability. This could result in a greater chance of not detecting a true alternative hypothesis or confidence intervals that are too large to provide useful information.

6.109 **a)** H_0: The patient is ill (or "the patient should see a doctor"); H_a: The patient is healthy (or "the patient should not see a doctor"). A Type I error means a false negative: clearing a patient who should be referred to a doctor. A Type II error is a false positive: sending a healthy patient to the doctor. Note that some students may switch the null and alternative hypotheses. They may assume the patient is healthy and let the results of the test provide evidence that the patient should see a doctor. **b)** One might wish to lower the probability of a false negative so that most ill patients are treated. On the other hand, if money is an issue, or there is concern about sending too many patients to see the doctor, lowering the probability of false positives might be desirable.

6.111 **a)** See the table below. **b)** See the graph that follows. **c)** The graph shows that as age increases, the months employed also increases. The width of the confidence intervals is the same for each of the ages.

Age	18	19	20	21	22	23	24	25	26
Months Employed, $\bar{x}$	2.9	4.2	5.0	5.3	6.4	7.4	8.5	8.9	9.3
95% CI, $\bar{x} \pm 0.32$	(2.58, 3.22)	(3.88, 4.52)	(4.68, 5.32)	(4.98, 5.62)	(6.08, 6.72)	(7.08, 7.72)	(8.18, 8.82)	(8.58, 9.22)	(8.98, 9.62)

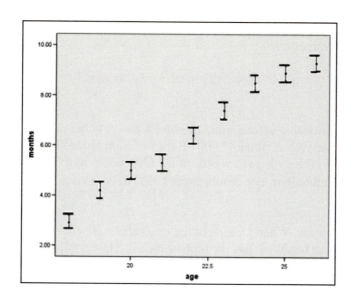

6.113   Applet.

6.115 $(-4.774, -3.806)$.

6.117   **a)**

2	0 3 4
2	
3	0 1 1 2 4
3	6
4	3

b) $(26.06, 34.74)$. **c)** Yes, because 25 is not in the interval, we can conclude that the average odor threshold of beginning students is higher.

6.119 $(\$959, \$968)$.

6.121   **a)** The population of interest would be all non-prescription medication consumers. The conclusions can be drawn for certain about the population of Indianapolis that has their phone number listed. **b)** Food stores: $(15.22, 22.12)$, mass merchandisers: $(27.77, 36.99)$, and pharmacies: $(43.68, 53.52)$. **c)** Yes, the confidence intervals do not overlap, which means the averages are different from each other.

6.123   **a)** With an increased sample size, the margin of error will decrease and the width of the interval decreases. **b)** As n increases, the sampling variability gets smaller and therefore the P-value decreases. **c)** As in part (b), as n increases, the sampling variability gets smaller and the power increases.

6.125   No, 0.05 is a probability about the sample mean, not about μ.

6.127   **a)** The probability of seeing this large a difference in the samples, if in fact the populations were not different, is very small. Therefore, the results of this study are considered to be significant. **b)** 95% confidence means that 95 out of 100 samples will provide an interval that is greater than zero. **c)** No, since the mothers voluntarily chose to participate, we cannot attribute a woman's being off welfare simply to training. There may be lurking variables involved.

6.129   As in the previous problem, one would expect to reject H_0 approximately 5 times out of 100.

6.131   **b)** With $n = 100$, this is a reasonable assumption. **c)** $m = 29.184$. **e)** One would expect approximately half of the intervals to contain 1000. Repeated simulations will not give the exact same results (unless the same seed was used for the random number generator). In a very large number of simulations, one would expect 50% to contain μ.

Case Study 6.1

For the 50- to 64-year age group: When looking at the category "ambience," the response "Tables are too close together" has the highest mean. This is followed by the restaurants being too noisy and the background music being too loud. The category "menu design" showed that the complaint with the highest mean was that the menu print was not large enough. "Service" showed that most people in this age group would rather be served than serve themselves.

The rank of the responses for the 65- to 79-year age group was the same as described for the 50- to 64-year age group.

The table below contains the averages for each response. The responses with an * indicate those that showed significant differences in means between the two age groups.

Question	50 to 64	65 to 79	z	P-value
Ambience				
Tables too close	3.79	3.81	−0.25	0.8026
Restaurants too loud*	3.27	3.55	−3.50	0.0005
Background music too noisy	3.33	3.43	−1.25	0.2113
Tables too small*	3.00	3.19	−2.38	0.0176
Too smoky	3.17	3.12	0.63	0.5320
Most restaurants are too dark*	2.75	2.93	−2.25	0.0244
Menu Design				
Print size not large	3.68	3.77	−1.13	0.2606
Glare*	2.81	3.01	−2.50	0.0124
Colors*	2.53	2.72	−2.38	0.0175
Service				
Want service rather than self-serve*	4.23	4.14	1.13	0.2606
Rather pay server than cashier*	3.88	3.48	5.00	5.74E-07
Service too slow	3.13	3.10	0.38	0.7077
Hard to hear*	2.65	3.00	−4.38	1.22E-05

Case Study 6.2

The responses with an asterisk (*) are those that had a significant difference in means between the two age groups.

	50 to 64	65 to 79	z	*P*-value
Accessibility and Comfort Inside				
Salad bars/buffets difficult to reach	3.04	3.09	−0.63	0.5320
Aisles too narrow*	3.04	3.20	−2.00	0.0455
Bench seats are too narrow*	3.03	3.25	−2.75	0.0060
Floors around bars/buffets often slippery*	2.84	3.01	−2.13	0.0336
Bathroom stalls too narrow*	2.82	3.10	−3.50	0.0005
Serving myself is difficult*	2.58	2.75	−2.13	0.0336
Most chairs are too small	2.49	2.56	−0.88	0.3816
Outside Accessibility				
Parking lots too dark at night*	2.84	3.26	−5.25	1.52E-07
Parking spaces too narrow*	2.83	3.16	−4.13	3.71E-05
Curbs near entrance difficult*	2.54	3.07	−6.63	3.49E-11
Doors too heavy*	2.51	3.01	−6.25	4.12E-10
Distance from parking lot too far*	2.33	2.64	−3.88	0.0001

Chapter 7: Inference for Distributions

7.1   **a)** 19.5. **b)** 15.

7.3   ($466.45, $549.55).

7.5 H_0: $\mu = \$550$, H_a: $\mu > \$550$, $t = -2.15$, $0.95 < P$-value < 0.975 using the t-table with 15 degrees of freedom. Based on this P-value, there is not enough evidence to believe that average apartment rents are greater than $550.

7.7   **a)** H_0: $\mu = 0$; H_a: $\mu \neq 0$. There is no specific reason to expect the change in sales to have increased or decreased, so a two-sided alternative hypothesis is appropriate. **b)** Using df = 30 to be conservative, $t = -1.687$, $0.10 < P$-value < 0.20. Do not reject the null hypothesis. **c)** No, this result tells us about the average, not the sales in every store. A positive change or a zero change would certainly be reasonable given this data.

7.9   SPSS gives a P-value of 0.024.

7.11   The confidence interval for the difference between the two drinks (drink B – drink A) is (–7.7440, 5.7440).

7.13   Yes, the sample size is greater than 40, so the t procedure would be appropriate.

7.15   Power = 0.9678.

7.17   **a)** 2.179. **b)** 2.060. **c)** 2.787. **d)** As the sample size increases, $t*$ decreases for the same confidence level. As the confidence level increases, the $t*$ also increases for the same sample size.

7.19   **a)** 17. **b)** 1.740 and 2.110. **c)** 0.05 and 0.025. **d)** $0.025 < P$-value < 0.05. **e)** Yes, significant at 0.05 but not significant at 0.01. **f)** Excel gives the P-value as 0.042421.

7.21   **a)** 149, but use df = 100 on Table D to be conservative. **b)** P-value < 0.0005. **c)** Excel (using 149 df) gives a P-value of 9.4145 x 10^{-5}.

7.23   (25.57, 32.03).

7.25   The confidence interval is (6.50, 11.46), using 100 degrees of freedom to be conservative.

7.27   **a)**

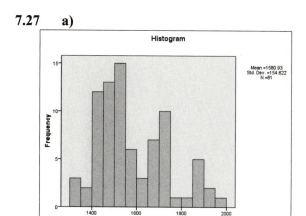

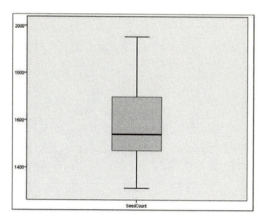

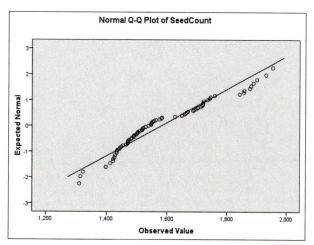

b) Based on the histogram and the boxplot, we can see that the distribution is somewhat skewed to the right. The Normal probability plot confirms that the distribution is not Normal based on the pattern of the dots. **c)** The robustness guidelines indicate that with a sample size of 81, the *t* procedures are appropriate even when the distribution is skewed.

7.29   **a)** H_0: $\mu = 1550$ and H_a: $\mu > 1550$. The test statistic is $t = 1.800$ with a *P*-value of 0.038. There is evidence that the average number of seeds in a one-pound scoop is greater than 1550. **(b)** H_0: $\mu = 1560$ and H_a: $\mu > 1560$. The test statistic is $t = 1.218$ with a *P*-value of 0.1135. There is not enough evidence that the average number of seeds in a one-pound scoop is greater than 1560. **(c)** With a 90% confidence interval of (1552.34, 1609.52), we would expect the true population mean to be less than 1552.34 approximately 5% of the time. Since 1550 is below the lower bound of this interval, we would expect a value like this to happen less than 5% of the time. However, 1560 is not below this lower bound, so we would expect a value this low or lower to occur more than 5% of the time.

7.31   **a)** The histogram of this data shows a skewed left distribution. The boxplot below confirms that there are both high and low outliers. The Normal probability plot shows that this is not a Normally distributed set of data. The *t* procedure is robust when the sample size is large, so we should still see reasonably accurate results even though the distribution is not Normal.

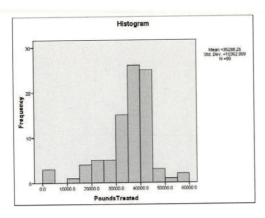

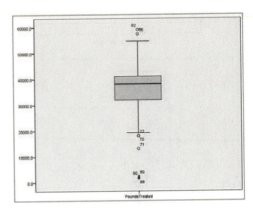

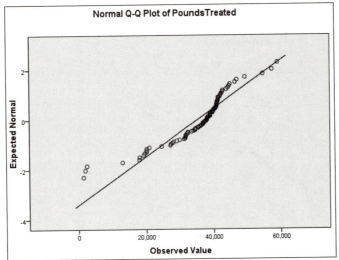

b) Mean = 35,288.25, standard deviation = 10,362.01, the standard error = 1092.25, and the margin of error = 1817.51. **c)** (33470.74, 37105.76). **d)** H_0: μ = 33000; H_a: $\mu >$ 33000. The P-value is 0.0195. There is evidence that the average pounds of product treated in one hour is greater than 33,000.

7.33 **a)** Six streams are classified very poor or poor out of 49 total, so the sample proportion is 0.122. **b)** No, sample proportions use Z^*, not t^*. Sample proportions are for categorical data, sample means are for quantitative data. We have converted quantitative data into categorical data by sorting the streams into very poor, poor, and other.

7.35 **a)** See the stemplot and Normal quantile plot below. The data show a distribution heavily skewed to the right with three high outliers. **b)** $\bar{x}$ = 31.134, s = 21.248, se = 3.360. **c)** (27.338, 40.930).

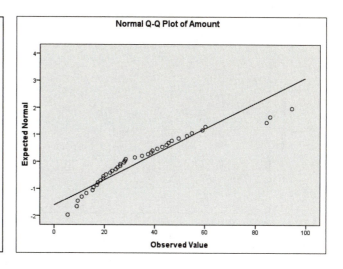

```
Amount Stem-and-Leaf Plot

 Frequency    Stem &  Leaf

     3.00       0 .  589
     9.00       1 .  025577899
    10.00       2 .  0234566788
     5.00       3 .  24789
     6.00       4 .  124569
     3.00       5 .  249
     1.00       6 .  0
     3.00  Extremes    (>=84)


Stem width:    10.00
Each leaf:      1 case(s)
```

7.37 3.64 ± 0.175.

7.39 (3.69, 3.81).

7.41 **a)**

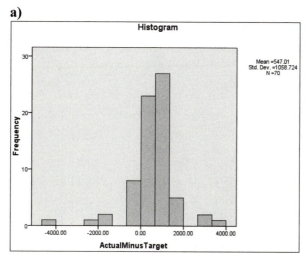

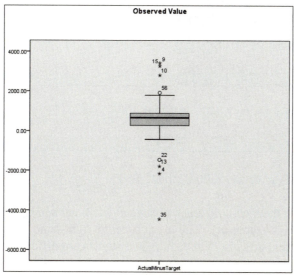

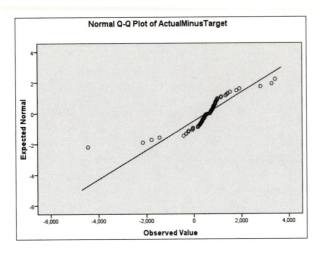

17.14% of the bags fall below the advertised target. **b)** The data are slightly skewed to the left. There are quite a few outliers, and the data are not Normal. With 70 observations in the data set, the *t* procedures will still give approximately correct results, even though Normality is not met. **c)** H_0: $\mu_{\text{diff}} = 500$, H_a: $\mu_{\text{diff}} < 500$. The test statistic is *t* $= 0.372$, and the *P*-value $= 0.6445$. There is not enough evidence that average difference between the number of seeds in the bag and the target number of seeds is less than 500. **d)** With this value excluded, the *P*-value of this test would increase as this would cause the test statistic to increase and the area below this test statistic to increase accordingly.

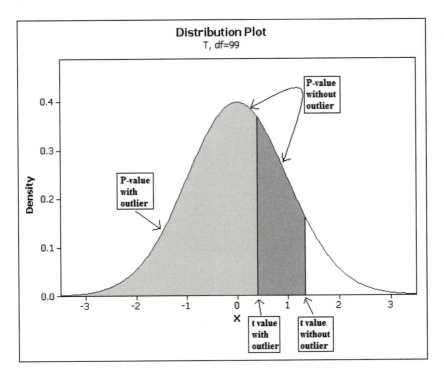

e) The *P*-value for H_0: $\mu_{\text{diff}} = 500$, H_a: $\mu_{\text{diff}} < 500$ is now 0.8695.

7.43   The 90% confidence interval for the mean difference between left-hand and right-hand thread times is (5.47 seconds, 21.17 seconds). The mean time for right-hand threads is 89% of the mean time for left-hand threads. This could be a significant difference in an assembly line. A 10% reduction in time will accumulate if the task is performed repeatedly throughout a day.

7.45   **(a)** The hypotheses are H_0: $\mu_{\text{diff}} = 0$ and H_a: $\mu_{\text{diff}} \neq 0$.   **(b)** The test statistic is $t = 4.358$ with a P-value of 0. There is evidence that there is a difference between the driver's calculations and the car's computer estimates of miles per gallon.

7.47   Taking the differences (variety A – variety B) to determine if there is evidence that variety A has the higher yield corresponds to the hypotheses H_0: $\mu = 0$, H_a: $\mu > 0$ for the mean of the differences. $t = 1.414$ with 9 degrees of freedom, and $0.05 < P$-value < 0.10. The sample data do not give overwhelming evidence in support of variety A having a higher mean yield than variety B.

7.49   **a)** One-sample design.   **b)** The best answer is 2 independent samples. Even though we are looking at changes of opinion over time from the same basic group, problems with missing values for one survey due to nonresponse or customers leaving the pool would make the paired sample difficult. However, if the exact same sample was used both years with a stable population of customers, matched pairs would also be an acceptable answer.

7.51   **a)** $t^* = 2.403$ from Table D using df $= 50$ to be conservative.   **b)** Reject H_0 when $t > 2.403$ or reject H_0 when $\bar{x} > 287.89$.   **c)** $P(\bar{x} > 287.89 \mid \mu = 300) = P(z > -0.1011) = 0.5398$. The bank should include more customers in the study.

7.53   **a)** H_0: median $= 0$, H_a: median > 0 or H_0: $p = 0.5$ H_a: $p > 0.5$.   **b)** Taking the left-hand time minus the right-hand time we count up the number of positives values (these correspond to right-hand times being faster than left-hand times). Using the Normal approximation to the binomial, $P(X \geq 19) = P(Z \geq 2.86) = 0.0021$. From software, we can find that $P(X \geq 19) = 0.0033$. These data show evidence against the null hypothesis in favor of the median difference being greater than 0.

7.55 H_0: median $= 0$; H_a: median > 0 or H_0: $p = 0.5$; H_a: $p > 0.5$. The probability that Jocko's estimate would be higher that the "trusted" garage's estimate at least 8 times out of 10 if the null hypothesis is true would be $P(X \geq 8) = 0.0547$. This is not very likely, but we would not reject the null hypothesis at the 5% level. In Exercise 7.44, the results were strong enough to conclude that Jocko's was giving higher estimates, but the results here were not.

7.57   **a)** A two-sided test makes more sense. We don't have information to tell us if one design should be better than another.   **b)** df $= 44$, but use 40 on Table D.   **c)** A t statistic above 2.021 or below -2.021 will lead to a decision to reject the null hypothesis.

7.59   The individual observations are no longer independent of each other, by design. You must compare individual sales for each day of the week.

7.61   By assigning the next ten employees to a new monitor and the following ten to a standard monitor, there may be influences due to time differences. At the end of the test period, the new monitor users will all have greater usage times. There may also be influences simply from employees talking to each other about their impressions.

7.63

Software	Excel	Minitab	JMP	SAS	SPSS
Means	July = 7.1075 Sept = 7.4235	July = 7.107 Sept = 7.423	Sept – July = 0.316000	July = 7.1075 Sept = 7.4235	July = 7.107500 Sept = 7.423500
Variability	Variance = 0.040478125 and 0.014970625	StDev = 0.201 and 0.122	Std Err Diff = 0.105308	Std. deviation = 0.2012 and 0.1224	Std. deviation = 0.2011918 and 0.1223545
Test Statistic	$t = -3.00072506$	$T = -3.00$	T ratio = 3.000725	t value = -3.00	$t = -3.001$
DF	7	6	6.602742	8	8
P-value	1 tail = 0.009960842, 2 tail = 0.019921684	$P = 0.024$	1 tail = 0.0107 2 tail = 0.0214	$\Pr > |t| =$ 0.0171	2 tail = 0.017
CI?	No	Yes	Yes	Yes	Yes

7.65 $t = 5.678$, df = 133, and P-value < 0.001. The test shows significant results for the difference in wheat prices between September and July.

7.67   Assume $n_1 = n_2 = n$.

Show that $s_p^{\,2}\left(\dfrac{1}{n}+\dfrac{1}{n}\right)=\dfrac{s_1^{\,2}}{n}+\dfrac{s_2^{\,2}}{n}$.

$$s_p^{\,2}=\frac{(n-1)s_1^{\,2}+(n-1)s_2^{\,2}}{n+n-2}=\frac{(n-1)(s_1^{\,2}+s_2^{\,2})}{2(n-1)}=\frac{1}{2}(s_1^{\,2}+s_2^{\,2})$$

$$\frac{1}{2}(s_1^{\,2}+s_2^{\,2})\left(\frac{1}{n}+\frac{1}{n}\right)=\frac{1}{n}(s_1^{\,2}+s_2^{\,2})=\frac{s_1^{\,2}}{n}+\frac{s_2^{\,2}}{n}.$$

7.69   **a)** Reject the null hypothesis because 0 is not inside the confidence interval. $H_0 : \mu_A = \mu_B$ can also be written as $H_0 : \mu_A - \mu_B = 0$. **b)** As sample size increases, the margin of error decreases.

7.71   The 95% confidence interval is (5.35, 14.65) using 40 degrees of freedom as the conservative estimate from the t table. A 99% CI will be wider than a 95% CI because the t^* value increases as the confidence level increases.

7.73   **a)** Yes it is appropriate to use the t procedures for these data since the sample sizes are large. **b)** $H_0 : \mu_F = \mu_M$; $H_a : \mu_F \neq \mu_M$. **c)** $t = 0.2760$, df = 36, P-value > 0.50. There is not enough evidence of a difference between the total cholesterol levels of males and females. **d)** (–12.0957, 15.8757). This interval shows that the females could have total cholesterol levels lower or higher than the males which agrees with the conclusion from the hypothesis test. **e)** If the students had modified their behavior based on taking the health class, such as eating healthier or exercising regularly, then the results might not generalize to the rest of the population. However, since we know these students were sedentary, they were likely not adjusting their activities based on the health class information, and the information would not be affected by knowing that these students were taking a health class.

7.75 **a)**

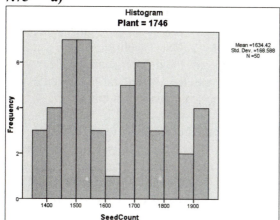

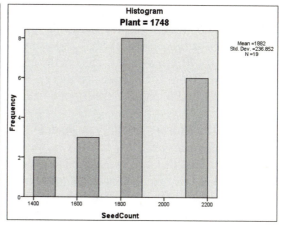

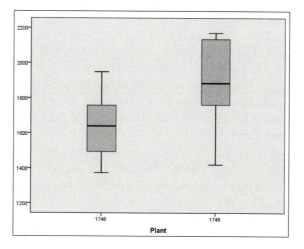

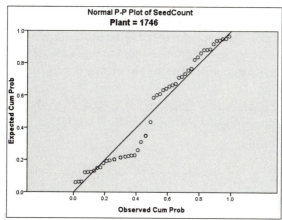

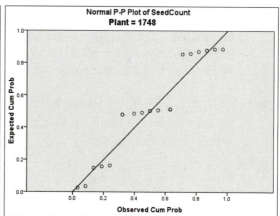

b) The sum of the sample sizes is 69, which is large enough to use the *t* procedures even though the data is not Normally distributed. **c)** The confidence interval is (−412.845,

–82.315) using software. **d)** $H_0: \mu_{1746} - \mu_{1748} = 0$; $H_a: \mu_{1746} - \mu_{1748} \neq 0$, $t = -4.172$, P-value ≈ 0. **e)** Based on the t procedures above, there is evidence that there is a population average difference in the seed counts for plant 1746 and plant 1748.

7.77   **a)** The means were 4.079 for females and 3.833 for males. **b)** The standard deviations were 0.986 for females and 1.068 for males. The t procedures are appropriate for these large sample sizes. **c)** $H_0: \mu_F = \mu_M$ and $H_a: \mu_F \neq \mu_M$. $t = 2.898$. P-value = 0.004. **d)** (0.079, 0.414). **e)** 0.25 is in the confidence interval for the difference in the means. While there is evidence that there is a difference between the average ratings given by males and females, there is not evidence of a difference of at least 0.25 units.

7.79   **a)** The data are probably not exactly Normally distributed because there are only five discrete answer choices. **b)** Yes, the large sample sizes would compensate for uneven distributions. The two-sample comparison of means t test is fairly robust. **c)** $H_0: \mu_I = \mu_C$ and $H_a: \mu_I > \mu_C$. (A "$\neq$" would also be appropriate in the alternative hypothesis, but be sure to double the P-value when checking your answer with this one.) **d)** $t = 3.57$, P-value < 0.0005. Reject the null hypothesis. There is strong evidence the average self-efficacy score for the intervention group is significantly higher than the average self-efficacy score for the control group. **e)** (0.191, 0.669).

7.81   **a)** See Exercise 7.80. **b)** (4.374, 5.426). **c)** $H_0: \mu_{D/B} = \mu_C$ and $H_a: \mu_{D/B} > \mu_C$. See the explanation in Exercise 7.80. $t = 18.49$, P-value is close to 0. Reject the null hypothesis. There is very strong evidence the average exposure to respirable dust is significantly higher for the drill and blast workers than it is for the concrete workers. **d)** The sample sizes are so large that a little skewness will not affect the results of the two-sample comparison of means test.

7.83   **a)** No. If we use the 68-95-99.7% rule, then 68% of the younger kids would consume between –2.5 and 18.9 ounces of sweetened drinks every day. The same problem with negative consumption for the older kids (starting with 95%) as well. **b)** $H_0: \mu_{older} = \mu_{younger}$ and $H_a: \mu_{older} \neq \mu_{younger}$. $t = 1.44$, $0.2 < P$-value < 0.3. Do not reject the null hypothesis. There is not enough evidence to say that there is a significant difference between the average consumption of sugary drinks between the older and younger groups of children. **c)** Using df = 4, $t^* = 2.776$, (–5.85, 18.45). **d)** The sample sizes are very uneven and fairly small. The sample data are not Normally distributed either. The t procedures are not particularly appropriate here. **e)** How were these children selected to participate? Were they chosen because they consume such large quantities of sweetened drinks?

7.85   **a)** $H_0: \mu_1 - \mu_2 = 0$; $H_a: \mu_1 - \mu_2 > 0$. $t = 22.16$, degrees of freedom are 1, $0.01 < P$-value < 0.02. Based on this small P-value, the conclusion is that the bread loses vitamin C after several days after baking. **b)** 26.91 ± 7.665 mg.

7.87   **a)** $H_0: \mu_1 - \mu_2 = 0$, $H_a: \mu_1 - \mu_2 > 0$, $t = -0.32$, df = 1, P-value > 0.25. There is no evidence that bread loses vitamin E several days after baking. **b)** -0.55 ± 10.73.

7.89   **a)** $H_0: \mu_1 - \mu_2 = 0$, $H_a: \mu_1 - \mu_2 \neq 0$, $t = -8.238$, P-value < 0.0005. The conclusion is that there is a significant difference in mean ego strength between the low fitness and high fitness groups. **b)** This is an observational study, not a designed experiment. There may be several lurking variables. **c)** Again, the lurking variables may be the cause of ego strength. Middle-aged men who think highly of themselves may be more disciplined and have a strong physical regimen than men who do not think highly of themselves.

7.91 **a)** (−0.91, 6.91) using 50 df on Table D. **b)** It is possible that there was actually a drop in average sales between the last year and this year. The data describe a sample of stores, not all stores in the chain.

7.93 **a)** The histograms that follow show that the distributions are somewhat skewed to the right for both men and women. However, as seen in the boxplots, there are no outliers in the distributions. Since the sum of the sample sizes is 38, and since neither strong skewness nor outliers are present, the *t* procedures are appropriate for these data.

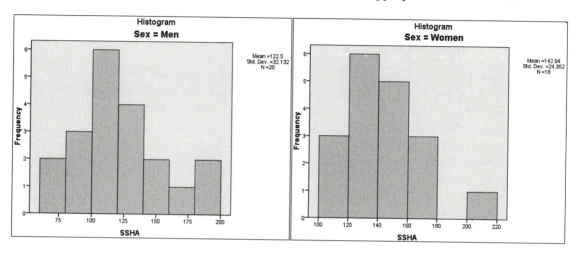

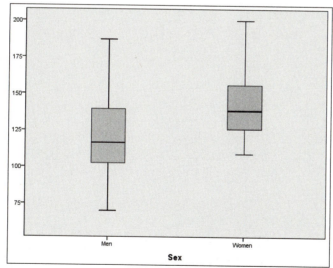

b) H_0: $\mu_{\text{women}} - \mu_{\text{men}} = 0$, H_a: $\mu_{\text{women}} - \mu_{\text{men}} > 0$, $t = 2.223$, *P*-value = 0.0165. There is evidence that women do have higher SSHA scores than men on average.
c) (4.907, 35.981).

7.95 Verify.

7.97 **a)** $t^* = 2.776$. **b)** $t^* = 2.306$. **c)** With the pooled procedure, it is easier to see a significant result in the sample data.

7.99   **a)** $F^* = 2.59$. **b)** For a two-sided test, this value is significant at the 10% level but not at the 5% level.

7.101   The power would decrease because you would be less likely to detect deviations from the null hypothesis when the difference in the means is smaller.

7.103   **a)** $F = 3.59$. **b)** 3.77. **c)** Do not reject the null hypothesis. There is no evidence that there is a difference in the two population standard deviations.

7.105   **a)** $F^* = 647.79$. The power is extremely low for unequal variances. **b)** $F = 3.96$. Fail to reject the null hypothesis.

7.107   **a)** H_0: $\sigma_1 = \sigma_2$; H_a: $\sigma_1 > \sigma_2$. **b)** $F = 1.74$. **c)** P-value > 0.10 using $F(19, 15)$ on Table E. The results of this sample are not significant.

7.109   The power for $n = 25$ is 0.966003; $n = 30$ is 0.986689; $n = 35$ is 0.994990; $n = 40$ is 0.998175. These powers were calculated using Minitab. The powers obtained using the Normal approximation are quite close to these values. As the number of scoops increases, the power also increases.

7.111   **a)** Power $= 0.5948$. **b)** Power $= 0.7794$.

7.113   **a)**

```
1 | 01233344
1 | 5566667778999999
2 | 00124444
2 | 5555566667
3 | 244
3 | 5
4 | 1
4 | 8
5 |
5 |
6 | 3
6 |
7 |
7 | 9
```

b) 23.56 ± 3.58.

7.115   **a)** The two-sided P-value is 0.08, which makes the one-sided P-value equal to 0.04. Since the test statistic is negative, the one-sided P-value is for the "less than" alternative hypothesis. So we would reject the null hypothesis. **b)** Since the test statistic is positive, we know that the area under the t curve to the right of this test statistic is 0.04, but we are testing a "less than" alternative hypothesis, so the P-value would be $1 - 0.04 = 0.96$.

7.117   Consumption of wine and food tends to be different when one eats alone rather than in a group. Since the entire table was assigned a wine label, this effect could be magnified. This study should take these items into account in order to have reliable results.

7.119 There is a significant difference between the high- and low-performing restaurants with regard to food serving in promised time, staff being well-dressed, serving ordered food accurately, and employees knowing the menu. There is not a significant difference in the other qualities. We need to assume fairly Normally distributed data without outliers and that a simple random sample was taken from each group.

Perceived Quality	$\bar{x}_H - \bar{x}_L$	$\sqrt{\dfrac{s_H^2}{170} + \dfrac{s_L^2}{224}}$	t	P-value	Conclusion
Food served in promised time	0.45	0.139	3.24	Between 0.001 and 0.002	Reject H_0
Quickly corrects mistakes	0.16	0.125	1.28	Between 0.2 and 0.3	Do not reject H_0
Well-dressed staff	0.39	0.136	2.87	≈ 0.005	Reject H_0
Attractive menu	0.21	0.143	1.47	Between 0.1 and 0.2	Do not reject H_0
Serving accurately	0.37	0.123	3.01	Between 0.002 and 0.005	Reject H_0
Well-trained personnel	0.06	0.125	0.48	> 0.5	Do not reject H_0
Clean dining area	0.08	0.127	0.630	> 0.5	Do not reject H_0
Employees adjust to needs	0.14	0.123	1.14	Between 0.2 and 0.3	Do not reject H_0
Employees know menu	0.29	0.119	2.44	Between 0.01 and 0.02	Reject H_0
Convenient hours	0.24	0.129	1.86	Between 0.05 and 0.1	Do not reject H_0

7.121 **a)** Study was done in South Korea; results may not apply to other countries. Only selected QSRs were studied; results may not apply to other QSRs. Response rate is low (394/950); we would trust the results more if the rate were higher. The fact that no differences were found when the demographics of this study were compared with the demographics of similar studies suggests that we do not have a serious problem with bias based on these characteristics. **b)** Answers will vary.

7.123 H_0: $\mu = 4.88$; H_a: $\mu > 4.88$, $t = 21.98$, P-value is close to 0, so we can reject the null hypothesis. There is strong evidence that hotel managers have a significantly higher average masculinity score than the general male population.

7.125 **a)** No outliers, slightly skewed right but fairly symmetric. See the stemplot below.
b) (12.9998, 13.3077).

```
12 | 22234
12 | 5556777888888899
13 | 01111223344444
13 | 5556677788
14 | 113
```

7.127 $\bar{x}_C = 48.9513$, $s_C = 0.21537$, $\bar{x}_R = 41.6488$, $s_R = 0.39219$. The side-by-side boxplot shows cotton much higher than ramie. H_0: $\mu_C = \mu_R$; H_a: $\mu_C > \mu_R$. $t = 46.162$, P-value is very close to 0 so reject the null hypothesis. There is strong evidence that cotton has a significantly higher mean lightness than ramie.

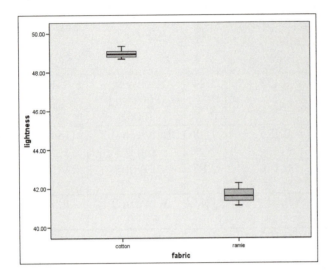

7.129   (1.0765, 7.6035) if equal variances not assumed; see SPSS output above.

7.131   (5.7403, 6.0270).

7.133   The 95% confidence interval on percentage of lower priced products at the alternate supplier is (64.55, 92.09). This suggests that more than half of the products at the alternate supplier are priced lower than the original supplier.

7.135   **a)** H_0: $\mu_1 = \mu_2$, H_a: $\mu_1 < \mu_2$, $t = -8.954$, P-value ≈ 0. Conclude that the workers were faster than the students. **b)** The t procedures are robust for large sample sizes even when the distributions are slightly skewed. **c)** The middle 95% of scores would be from 29.66 to 44.98. **d)** The scores for the first minute are clearly much lower than the scores for the 15th minute.

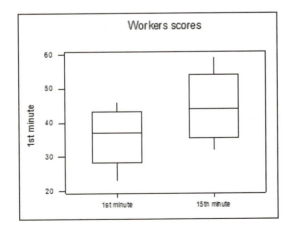

7.137   **a)** (0.207, 0.573). **b)** (−0.312, 0.012). **c)** We are 90% confident that the average daily alcohol consumption of London double-decker bus conductors is between 0.207 and

0.573 grams of alcohol. We are 80% confident that, on average, drivers consume between 0.312 grams less and 0.012 grams more alcohol per day than conductors.

7.139 As the degrees of freedom increase for small values of n, the value of t^* rapidly approaches the z^* value. As sample sizes get larger, the t^* values are close to $z^* = 1.96$ but never become greater than 1.96.

Case Study 7.1

	Mean	St. Dev.	95% CI	Five-number Summary
2002 billings	5.59	5.45	(3.109, 8.071)	1.6, 2, 2.9, 7.85, 24.8
2001 billings	5.49	5.32	(3.070, 7.911)	1.1, 1.75, 3.1, 8.1, 23.7
Architects	10.57	8.90	(6.519, 14.624)	1, 5, 7, 16.5, 39
Engineers	6.81	10.80	(1.895, 11.724)	0, 0, 2, 9, 36
Staff	59.90	55.89	(34.464, 85.346)	9, 15.5, 58, 71, 240

Group Statistics

	Old = Pre1970 New = 1970 or later	N	Mean	Standard Deviation	Std. Error Mean
Bill02	New	12	6.200	6.629	1.914
	Old	9	4.778	3.540	1.180
Bill01	New	12	5.967	6.483	1.871
	Old	9	4.856	3.480	1.160
Architects	New	12	11.25	6.483	3.139
	Old	9	9.67	5.831	1.944
Engineers	New	12	6.42	10.317	2.978
	Old	9	7.33	12.021	4.007
Staff	New	12	57.08	63.484	18.326
	Old	9	63.67	47.323	15.774

Since the pooled procedures require that the two populations have equal variances, this is probably a safe assumption for architects and engineers but not for Bill02, Bill01, and staff, based on the differences in the standard deviations in the samples for these variables. We also need to assume that the "new" and "old" groups are independent. (For example, none of the "new" firms are branches of the "old" firms.)

The results of the two-sample comparison of means tests (new – old) from SPSS follow. There is not a significant difference for any of these variables when we compare old versus new companies.

Independent Samples Test

		Levene's Test for Equality of Variances		t-test for Equality of Means						
		F	Sig.	t	df	Sig. (2-tailed)	Mean Difference	Std. Error Difference	95% Confidence Interval of the Difference	
									Lower	Upper
ArchBill02	Equal variances assumed	1.023	.325	-.582	19	.567	-1.4222	2.4439	-6.5373	3.6928
	Equal variances not assumed			-.633	17.481	.535	-1.4222	2.2481	-6.1554	3.3110
ArchBill01	Equal variances assumed	1.128	.301	-.464	19	.648	-1.1111	2.3922	-6.1180	3.8958
	Equal variances not assumed			-.505	17.521	.620	-1.1111	2.2018	-5.7460	3.5238
Architects	Equal variances assumed	2.186	.156	-.395	19	.697	-1.583	4.011	-9.979	6.812
	Equal variances not assumed			-.429	17.513	.673	-1.583	3.692	-9.355	6.188
Engineers	Equal variances assumed	.637	.435	.188	19	.853	.917	4.880	-9.297	11.130
	Equal variances not assumed			.184	15.779	.857	.917	4.993	-9.679	11.513
Staff	Equal variances assumed	.066	.800	.261	19	.797	6.583	25.240	-46.244	59.411
	Equal variances not assumed			.272	18.998	.788	6.583	24.180	-44.027	57.194

Note that for all variables, BSA is a high outlier for the new firms. Also note that for architects, ArchBill01, and ArchBill02, CSO and Browning are high outliers for the old firms; for staff and engineers, American is a high outlier for the old firms. Many of these boxplots show either skewness or outliers, and for groups with such small sample sizes (9 for the old firms, 12 for the new firms), that is a problem. The *t* procedures are not necessarily appropriate here.

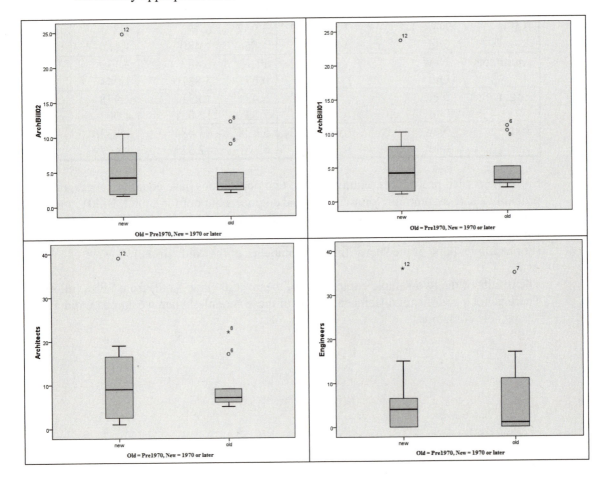

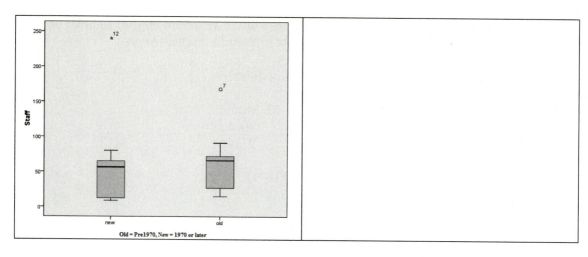

Case Study 7.2

Stem-and-Leaf of 4 BR $N = 9$
Leaf Unit = 10000

```
1 | 23
1 | 55
1 | 7
1 |
2 |
2 | 23
2 | 4
2 |
2 | 9
```

Stem-and-Leaf of 3 BR $N = 28$
Leaf Unit = 10000

```
0 | 6777
0 | 8899
1 | 01111
1 | 22222223
1 | 445
1 |
1 | 9
2 | 0
2 |
2 | 5
2 | 6
```

Both distributions appear skewed heavily to the right. It is more difficult to see this with the four-bedroom homes because the data set is very small. The Minitab output for the two-sample t test assuming unequal variances follows. It appears that there is strong evidence to indicate there is a difference in mean selling prices of four-bedroom and three-bedroom homes. A one-sided alternative would make sense because it is logical to expect that four-bedroom homes would sell for more than three-bedroom homes. While these are not SRSs, one might justify using the *t* test if we treated these as samples of homes from all future homes that will be sold in West Lafayette, Indiana.

Two-sample *T* for Four-Bedroom versus Three-Bedroom

	N	Mean	St. Dev.	SE Mean
4 BR	9	194944	57204	19068
3 BR	28	129546	49336	9324

95% CI for mu 4 BR – mu 3 BR: (19152, 111644)
T-test mu 4 BR = mu 3 BR (vs. not =): *T* = 3.08
P = 0.0095, df = 12

Chapter 8: Inference for Proportions

8.1 **a)** 780. **b)** 283 of the banks plan to acquire another bank. **c)** 0.3724.

8.3 **a)** 0.0175. **b)** 0.3724 ± 0.0344. **c)** (33.80%, 40.68%).

8.5 (0.50, 0.54). This is the same confidence interval seen in Example 3.

8.7 Answers may vary. One example is a 95% confidence interval, where $X = 15$ and $n = 30$. The confidence interval here is (0.321, 0.678). The plus four interval would be (0.332, 0.668). The differences will not be substantial when we hold to the assumption that the number of successes and number of failures are both at least 15.

8.9 (44.10%, 85.90%). We can say with 95% confidence that the population proportion of people who get better sun protection from your product is in this interval. The confidence interval gives an estimate for the population proportion, while the hypothesis test only gives information regarding the population proportion of 50%.

8.11 $z = 1.90$ with a P-value of 0.0574. The null hypothesis would still not be rejected at the 5% level.

8.13 Since there are five choices and two are recorded as "yes," use a p^* of 0.4, which assumes each choice is equally likely. This results in a necessary sample size of 93 people.

8.15 **a)** Hypotheses need population parameters, not sample statistics. H_0: $p = 0.3$ is appropriate. **b)** The Z test statistic should be used, not t, for population proportion significance tests. **c)** The standard error needs to be multiplied by z^* before adding and subtracting from $\hat{p}$.

8.17 **a)** $\hat{p} \sim N(0.4, 0.0693)$. **b)** See graph below. **c)** $p^* = 0.1386$.

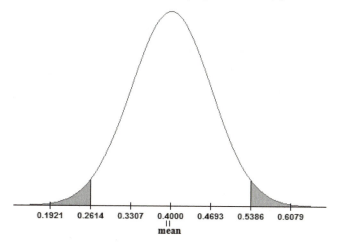

8.19 **a)** $\hat{p} = 0.161$; (0.1508, 0.1712). **b)** $\hat{p} = 16.1\%$; (15.08%, 17.12%).

8.21 **a)** The confidence interval would be (0.6412, 0.6988). **b)** The confidence interval would be (0.6534, 0.6866). **c)** While changing the size of the group that does not currently play an instrument does affect the confidence interval, it does not have a great effect. The margin of error is smaller when it is assumed that a larger portion of the group do not currently play a musical instrument.

8.23   **a)** 0.4171 ± 0.0164. **b)** The margin of error depends on the confidence level (the same for Exercise 8.22 as in Exercise 8.23), the proportion of successes (not the same for males and females in these stories), and the sample size (not the same for these stories).

8.25   **a)** 67,179. **b)** (41.68%, 42.32%).

8.27   **a)** (0.300, 0.354). **b)** (0.301, 0.355). The methods have the same margin of error = 0.027, but the plus four method shifts the interval slightly higher. **c)** Nonresponse rate is only 3.64%, which is small, so we can trust these results. **d)** Yes, the person delivering the sermon probably thinks the sermon is shorter than it actually is, and the congregation would probably think the sermons are longer than they actually are.

8.29   (0.7380, 0.7823).

8.31   Decreasing the confidence level will decrease the width of the confidence interval. The new interval will be (41.76%, 42.24%).

8.33   (0.635, 0.745).

8.35   (0.218, 0.251).

8.37   (0.766, 0.887).

8.39   (0.642, 0.768).

8.41 H_0: $p = 0.36$; H_a: $p \neq 0.36$. $Z = 0.9317$. *P*-value = 0.3524. There is no evidence to believe the sample does not represent the population with respect to rural versus urban residence.

8.43   At least 764 are needed in the sample.

8.45   **a)** H_0: $p = 0.5$; H_a: $p \neq 0.5$. $Z = 1.34$. *P*-value = 0.1802. There is no significant evidence that Kerrich's coin does not have probability 0.5 of coming up heads. **b)** (0.4969, 0.5165).

8.47   At least 289 are needed in the sample.

8.49   We need at least 201 in the sample; $m = 0.0691$.

8.51   See the table below.

$\hat{p}$	m
0.1	0.0480
0.2	0.0640
0.3	0.0733
0.4	0.0784
0.5	0.0800
0.6	0.0784
0.7	0.0733
0.8	0.0640
0.9	0.0480

8.53 Error! Bookmark not defined.$\mu_D = -0.1$; $\sigma_D = 0.1339$.

8.55 **a)** $\mu_{\hat{p}_1} = p_1, \mu_{\hat{p}_2} = p_2, \sigma_{\hat{p}_1} = \dfrac{p_1(1-p_1)}{n_1}, \sigma_{\hat{p}_2} = \dfrac{p_2(1-p_2)}{n_2}$. **b)** $\mu_D = \mu_{\hat{p}_1} - \mu_{\hat{p}_2}$.

   c) $\sigma_D{}^2 = \dfrac{p_1(1-p_1)}{n_1} + \dfrac{p_2(1-p_2)}{n_2}$.

8.57 **a)** $\hat{p}_m = 0.600$; $\hat{p}_w = 0.575$. **b)** $D = 0.025$. **c)** $SE_D = 0.1100$. **d)** 0.025 ± 0.216.

8.59 Plus four interval: $0.071 \pm 0.196 = (-0.125, 0.267)$. Z interval: $0.080 \pm 0.192 = (-0.112, 0.272)$. The plus four interval has a slightly lower sample proportion and a slightly larger margin of error.

8.61 **a)** H_0: $p_m = p_w$; H_a: $p_m \neq p_w$. This set of hypotheses has a two-sided alternative hypothesis since there is no information ahead of time to lead us to believe that one group would be more likely to lie about their height. **b)** The test statistic is $z = 1.12$ (or $z = -1.12$), giving a P-value of 0.2628. **c)** Do not reject H_0. There is not enough evidence of a difference in the population proportion of those who lied about their height for men and women.

8.63 **a)**

Population	Population Proportion	Sample Size	Count of Successes	Sample Proportion
Feb.–April 2006	p_1	2822	198	0.0702
May 2008	p_2	1553	295	0.1900

   b) The estimated difference is $D = 0.1198$ (or $D = -0.1198$ depending on the order of subtraction). **c)** Since there are more than 10 successes and more than 10 failures in each sample, it is appropriate. **d)** The confidence interval is $(0.0981, 0.1415)$ or $(-0.1415, -0.0981)$. **e)** $D = 11.98\%$; confidence interval: $(9.81\%, 14.15\%)$. **f)** The dates could definitely make a difference if more people tend to download podcasts at different times of the year. This could happen in May if people download more podcasts when school is not in session.

8.65 **a)** We can not conclude that the population number of downloads was 2.7 times as many as in the 2006 period since the proportion will vary between different samples from the same population. **b)** From the hypothesis test in Exercise 8.64, we can conclude that a different number of people downloaded podcasts. We could also do a one-sided hypothesis test which would show that a significantly greater proportion downloaded podcasts during the 2008 time period. To conclude that this applies to the whole years of 2006 and 2008, we would need to assume that the periods sampled from are representative of the entire year from which they were sampled.

8.67 H_0: $p_{adult} = p_{teen}$, H_a: $p_{adult} \neq p_{teen}$. $z = 17.88$. *P*-value = 0. There is evidence of a significant difference in the proportion of teens and the proportion of adults who played games on game consoles.

8.69 H_0: $p_{adult} = p_{teen}$, H_a: $p_{adult} \neq p_{teen}$. $z = 1.60$. *P*-value = 0.1096. There is no evidence of a difference in the proportion of teens and the proportion of adults who played games on computers.

8.71   **a)** $\mu_{\hat{p}_1 - \hat{p}_2} = -0.1$ and $\sigma_{\hat{p}_1 - \hat{p}_2} = 0.0947$. **b)** See the sketch that follows. **c)** Using the 68-95-99.7% rule tells us that this range is $\mu \pm 2\sigma = -0.1 \pm 0.189 = (-0.289, 0.089)$. See the sketch that follows.

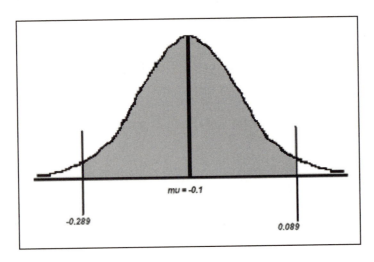

8.73   **a)** $\hat{p}_F \sim N(0.83, 0.0188)$, $\hat{p}_M \sim N(0.85, 0.0179)$. **b)** $(\hat{p}_M - \hat{p}_F) \sim N(0.02, 0.0259)$.

8.75   **a)** H_0: $p_1 = p_2$; H_a: $p_1 \neq p_2$. **b)** $Z = 1.22$, *P*-value = 0.2224. There is no evidence to suggest that there is a difference in tree preference between rural and urban populations. **c)** (−0.0209, 0.1389).

8.77   **a)** $\hat{p}_F = 0.800$, $\hat{p}_M = 0.394$. **b)** (0.275, 0.537). The data show that there is gender bias in the text because women are more often referred to with a juvenile reference. Zero is not included in the confidence interval, so there is a significant difference between the proportions of female and male juvenile references. Since both endpoints of the confidence interval are positive, women have a significantly higher proportion of juvenile references than men do.

8.79   **a)** $z = 0.73$. The *P*-value is 0.4654. Do not reject H_0. There is not enough evidence that the population proportion of students employed in the summer is significantly different for men and women. **b)** It is less likely that the difference in the population proportions will be detected when the sample sizes are smaller.

8.81   The 95% confidence interval is 0.11 ± 0.035.

8.83   **a)** The 95% confidence interval is 0.11 ± 0.039. H_0: $p_{2000} = p_{2004}$, H_a: $p_{2000} \neq p_{2004}$. The test statistic is $z = 5.48$, which still results in a P-value of approximately 0. The conclusion remains the same as in Exercise 8.82. **b)** The 95% confidence interval is 0.11 ± 0.033. H_0: $p_{2000} = p_{2004}$, H_a: $p_{2000} \neq p_{2004}$. The test statistic is $z = 6.59$, which still results in a P-value of approximately 0. There is evidence that the proportions are different in 2000 and 2004. **c)** In this case, there was no change to the conclusion based on the change in assumption of sample size. However, we can notice that the confidence interval was narrower with the larger sample size, and there was more evidence against the null hypothesis with this larger sample size as well. In some cases, this may result in drawing an incorrect conclusion if we don't know the correct sample size.

8.85   See a chart below of some potential sample sizes and confidence intervals. The margin of error decreases as the sample size increases and vice versa.

Sample Size	50	100	500	1430	2000	10000
Confidence Interval	(0.567, 0.822)	(0.605, 0.785)	(0.655, 0.735)	(0.671, 0.719)	(0.675, 0.715)	(0.686, 0.704)

8.87   **a)** Assuming that the events are independent (where the events are "the confidence interval contains the true proportion"), then the probability that two selected intervals would both contain the population proportion would be $0.95^2 = .9025$. So, the chance that one out of any two misses the population proportion would be 9.75%. The chance is even greater with 6 intervals. **b)** $z = 2.64$.

c)

Genre	Racing	Puzzle	Sports	Action	Adventure	Rhythm
Confidence Interval	(0.705, 0.775	(0.684, 0.756)	(0.643, 0.717)	(0.633, 0.707)	(0.622, 0.698)	(0.571, 0.649)

8.89   **a)** 0.164. 2460 households. **b)** (0.158, 0.17). **c)** $16.4\% \pm 0.6\%$. **d)** $D = 0.122$. **e)** 0.0067.

8.91 $H_0 : p_1 = p_2$, $H_a : p_1 \neq p_2$, $z = 20.18$, P-value is approximately 0, so reject the null hypothesis. There is evidence of a significant difference between the proportion of male athletes who admit to cheating and the proportion of female athletes who admit to cheating. The 95% confidence interval (male – female) is (0.1963, 0.2377). Someone who gambles will be less likely to respond to the survey. Do you think men or women are more likely to report that they do not gamble when, in fact, they do gamble?

8.93   **a)** and **b)**

Category	$\hat{p}$ (in %)	n	m (in %)
Download less	38	247	3.09
Peer-to-peer	33.3	247	3.00
E-mail and IM	24	247	2.72
Web sites	20	247	2.55
iTunes	17	247	2.39
Overall use of new services	7	1,371	0.69
Overall use paid services	3	1,371	0.46

c) Argument for (A): Readers should understand that the population percent is not guaranteed to be at the sample percent, there is variability involved in taking a sample. Argument for (B): Listing each individual margin of error does seem excessive, so you could summarize by saying that the margin of error was no greater than 3.09% for each of these questions. You could also separate out the last two questions by saying their margin of error was less than 1%.

8.95   **a)** $\hat{p}_{repeat} = 0.783$, $\hat{p}_{norepeat} = 0.517$, 95% CI (repeat – no repeat) is (0.102, 0.430).

b) $H_0: p_1 = p_2$, $H_a: p_1 \neq p_2$. $Z = 3.05$, P-value = 0.0022, so reject the null hypothesis. There is strong evidence of a significant difference in the proportion of tips received between servers who repeat the customer's order and those who do not repeat the order. **c)** Cultural differences, personalities of the servers, gender differences could all play a role in interpreting these results. Did one server only do the repeating while the other server did no repeating, or did they switch off? **d)** Answers will vary.

8.97 $H_0: p_1 = p_2$, $H_a: p_1 \neq p_2$, where p_1 represents the proportion of diehard fans that attend a Cubs game at least once a month. $z = 9.07$, P-value = 0. A 95% confidence interval on the difference is (0.3755, 0.5645). Diehard fans are more likely to attend games at least once a month than the less loyal fans.

8.99   The 95% confidence interval is (0.5524, 0.6736).

8.101 $H_0: p = 0.11$, $H_a: p < 0.11$, $z = -3.14$, P-value = 0.001. The results indicate there has been a significant decrease in the proportion of nonconformities. We are assuming the sample used is a SRS from the process.

8.103

n	10	25	50	100	150	200	400	500
m	0.438	0.277	0.196	0.139	0.113	0.098	0.069	0.062

As sample size increases, the margin of error decreases.

8.105 Starting with a sample size of 25 for the first sample, it is not possible to guarantee a margin of error of 0.15 or less. $m = 1.960\sqrt{\dfrac{(0.5)(0.5)}{25} + \dfrac{(0.5)(0.5)}{n_2}}$. This leads to a negative value for n_2, which is not feasible.

8.107 **a)** $p_0 = 0.791$. **b)** $\hat{p} = 0.3897$, $Z = -29.11$, P-value = 0. The proportion of Mexican-Americans on juries in this county is significantly lower than their proportion in the population. **c)** $z = -28.96$, P-value = 0. The results agree with the results in part (b).

Case Study 8.1

When comparing proportions of "girl" references to proportion of "boy" references, there does not appear to be any pattern that differentiates the male authors from the female authors. Texts 2, 3, 6, and 10 showed a significant difference between the two proportions (with P-value < 0.05) with proportion of "girl" references greater than proportion of "boy" references.

Case Study 8.2

$\hat{p}_1$	$\hat{p}_2$	n	z	*P*-value for Two-sided Test	Conclusion at 5% Significance Level
0.75	0.5	12	1.26	0.2059	Do not reject H_0.
0.75	0.5	20	1.63	0.1025	Do not reject H_0.
0.75	0.5	40	2.31	0.0209	Reject H_0.
0.75	0.5	80	3.27	0.0011	Reject H_0.
0.75	0.5	100	3.65	0.0003	Reject H_0.
0.75	0.5	200	5.16	≈ 0	Reject H_0.
0.75	0.5	500	8.16	≈ 0	Reject H_0.

As the sample size increases, the test statistic increases, the *P*-value decreases, and we are more likely to reject the null hypothesis.

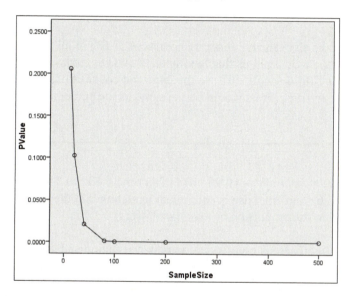

$\hat{p}_1$	$\hat{p}_2$	n	SE	Margin of Error	CI Lower Bound	CI Upper Bound
0.75	0.5	12	0.190941	0.374244	−0.12424	0.624244
0.75	0.5	20	0.147902	0.289888	−0.03989	0.539888
0.75	0.5	40	0.104583	0.204982	0.045018	0.454982
0.75	0.5	80	0.073951	0.144944	0.105056	0.394944
0.75	0.5	100	0.066144	0.129642	0.120358	0.379642
0.75	0.5	200	0.046771	0.091671	0.158329	0.341671
0.75	0.5	500	0.02958	0.057978	0.192022	0.307978

As the sample size increases, the margin of error decreases and the confidence interval narrows.

Chapter 9: Inference for Two-Way Tables

9.1   **a)** It would be best to consider gender as the explanatory variable since a person's gender may affect whether they would lie about their height. Whether a person lies about their height would not affect their gender. **b)** See part (d). There are two options for each variable, so r and c would both be 2. **c)** 4.

d)

	Men	Women	Total
Lied	22	17	39
Did Not Lie	18	23	41
Total	40	40	80

9.3   108/123 = 87.8% of successful firms and 34/47 = 72.3% of unsuccessful firms were offered exclusive territories.

9.5   There are 142 firms that have an exclusive-territory contract. 27.6% of all firms are unsuccessful. No association would tell us that having an exclusive-territory contract does not affect whether the firm is successful. So the expected count of unsuccessful firms among the exclusive-territory firms would be the same as the percent for all companies, 27.6%.

9.7 $df = (r - 1)(c - 1) = 5 \times 3 = 15$.

9.9   **a)** $\chi^2 = 10.95$. The value of $z^2 = (3.31)^2 = 10.95$. **b)** $(3.291)^2 = 10.83$. **c)** The statement of no relation between gender and label use is equivalent to stating that the proportion of female label users is equal to the proportion of male label users.

9.11   **a)** See Exercise 2.115. **b)** H_0: There is no association between region and whether or not a community bank offers RDC, H_a: There is an association between region and whether or not a community bank offers RDC. $\chi^2 = 24.059$, $df = 5$.
c) P-value = 0.

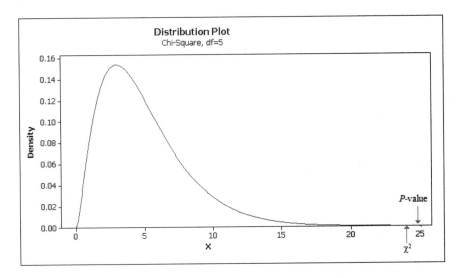

d) There is substantial evidence that there is an association between region and whether or not a community bank offers RDC.

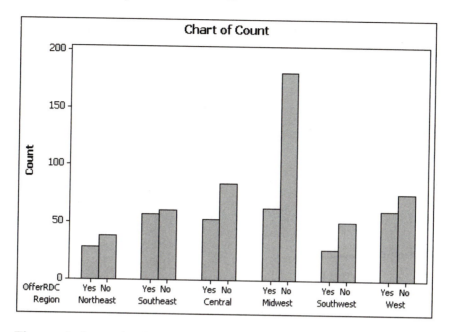

The graph shows clearly that the percent who offer and who don't offer RDC is almost the same for the Southeast. In the Central, Northeast, Southwest, and West, those who don't offer RDC make up a greater percent than those who do. In the Midwest, however, more than twice as many banks do not offer RDC than the number that do.

9.13   **a)** through **c)** See Exercise 2.120. **d)** H_0: There is no association between gender and whether or not a student lied to a teacher; H_a: There is an association between gender and whether or not a student lied to a teacher. $\chi^2 = 5351.94$, df = 1, *P*-value = 0. There is substantial evidence that there is an association between gender and whether or not a student lied to a teacher. A student is more likely to have lied to a teacher at least once if the student was a female. A chi-square graph with one degree of freedom is shown below. The test statistic would be so far to the right of the graph that a graph showing its location would show no other features of the graph.

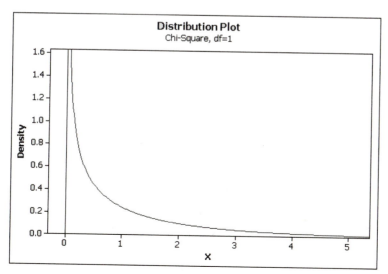

9.15   **a)** through **d)** See Exercise 2.123. **e)** The hypotheses are H_0: There is no association between age and whether an applicant was hired or not and H_a: There is an association between age and whether an applicant was hired or not. $\chi^2 = 9.106$, df $= 1$, P-value $= 0.003$. Reject the null hypothesis, as there is strong evidence of an association between age and whether or not an applicant was hired.

9.17   **a)** through **d)** See Exercise 2.128. **e)** $\chi^2 = 24.863$, P-value ≈ 0. There is strong evidence that there is an association between gender and whether or not a student is admitted to Wabash Tech. **f)** Business School: $\chi^2 = 10.390$, and the P-value is 0.001. We can conclude that there is an association between gender and whether or not the applicant is admitted for the business school. Law School: $\chi^2 = 20.481$, and the P-value is 0. We can conclude that there is even stronger evidence of an association between gender and whether or not the applicant is admitted for the law school.

9.19   **a)** V will have observations of zero 50% of the time when $U = 0$ or when $U = 1$. H_0: There is no association between U and V. When $a = 0$, the distribution of V will be the same regardless of the value of U.

b) and **c)**

a	0	5	10	15	20	25
% zeros for $V \mid U = 1$	50%	52.6%	55.6%	58.8%	62.5%	66.7%
χ^2	0.000	0.501	2.020	4.604	8.333	13.333
P-value	1.000	0.479	0.155	0.032	0.004	0.000

As the percent of zeros for V increases, the chi-square value increases, increasing at a greater rate the farther a gets away from zero. The P-value decreases as the percent of zeros for V increases.

d)

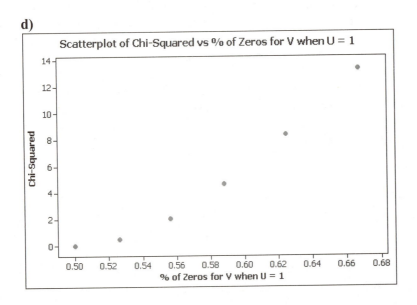

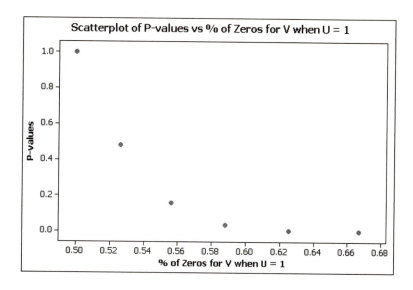

The chi-square value decreases as the *P*-value increases. When the percentages within the group change, the *P*-value reacts very quickly to the change. The chi-square value reacts a little less quickly. The greater the change in the percentages in the table, the greater the change in the chi-square statistic. The *P*–value also continues to decrease, but not as quickly as with the initial change.

9.21 **a)**

	Apr 2001	Apr 2004	Mar 2007	Apr 2008
Broadband	113	540	1,080	1,237
No Broadband	2,137	1,710	1,170	1,013

b) H_0: There is no association between the date of the survey and whether or not a home has broadband; H_a: There is an association between the date of the survey and whether or not a household accesses the Internet using broadband. $\chi^2 = 1599.503$, df = 3, *P*-value = 0. There is evidence that there is an association between the date of the survey and whether or not a household had broadband. **c)** $\chi^2 = 21.930$, df = 1. The *P*-value is still 0, so we would still reject the null hypothesis, but the result is not nearly as strong as that seen when all four dates were considered.

9.23 Answers will vary.

9.25 **a)**

	Claim Allowed	Claim Not Allowed	Total
Small	51	6	57
Medium	12	5	17
Large	4	1	5
Total	67	12	79

b) 10.53% of small claims, 29.41% of medium claims, and 20% of large claims were not allowed. **c)** H_0: There is no association between the size of the claim and whether or not the claim was allowed; H_a: There is an association between the size of the claim and whether or not the claim was allowed. **d)** $\chi^2 = 3.721$, df = 2, *P*-value = 0.156. There is not enough evidence to indicate that an association exists between the size of the claim and whether or not it was allowed.

9.27 $\chi^2 = 308.321$, df = 2 and the *P*-value = 0. There is evidence that the changes in the gateway course did have a significant impact on the DFW rate. Looking at the rates themselves, we can see that the DFW rates decreased with the changes that were implemented.

	DFW Count	ABC Count	Total
Year 1	1,019	1,389	2,408
Year 2	579	1,746	2,325
Year 3	423	1,703	2,126
Total	2,021	4,838	6,859

9.29   **a)**

	Entered After HS	Entered Later
Trades	320	622
Design	274	310
Health	2,034	3,051
Media / IT	976	2,172
Service	486	864
Other	1,173	1,082

b) Using a chi-square test of association with 5 degrees of freedom to determine whether there is an association between time of entry and field of study with 5 degrees of freedom, the test statistic is $\chi^2 = 276.110$, and the *P*-value is 0. **c)** There is evidence that there is an association between the field of study and the time of entry of Canadian students enrolled in private career colleges. A bar graph shows that those entering the media/IT field were the most likely to enter later, whereas those entering "other" fields and "design" were the groups most likely to enter right after leaving high school.

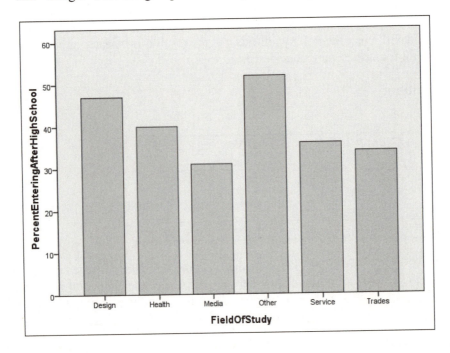

9.31 **a)**

	Family Funded	Not Family Funded
Trades	188	754
Design	222	377
Health	1,361	3,873
Media/IT	518	2,720
Service	248	1,130
Other	943	1,357

b) H_0: There is no association between field of study and whether or not the student received funding from family; H_a: There is an association between field of study and whether or not the student received funding from family. $\chi^2 = 544.787$, df = 5, *P*-value = 0. There is evidence that there is an association between the field of study and whether or not the student received funding from parents, spouse, or family. **c)** A bar graph shows that those entering the media/IT, service, or trade fields were the least likely to used funds from family while those entering "other" fields and "design" were the groups most likely to use family funding.

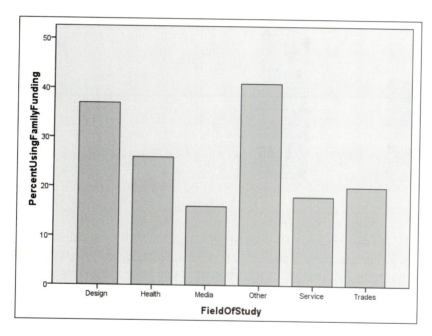

9.33 Answers will vary, but one example would be:

	X	Y	Z	Totals
A	10	10	10	30
B	10	10	10	30
C	10	10	10	30
Totals	30	30	30	90

9.35 Percent of department A's classes that are small: 32 / 52 = 61.54%. Percent of department B's classes that are small: 42 / 106 = 39.62%. Percent of department A's classes that are for third- and fourth-year students: 40 / 52 = 76.92%. Percent of department B's classes that are for third- and fourth-year students: 36 / 106 = 33.96%. department A teaches a much larger percentage of small classes, but department A teaches more than twice the percentage of third- and fourth-year students as department B.

9.37 $\chi^2 = 2.591$, *P*-value = 0.107 from SPSS. Do not reject the null hypothesis. There is not enough evidence to say that there is an association between model dress and magazine readership age group. The marginal percentages for model dress were 73.6% of the ads had models dressed not sexually and 26.4% had models dressed sexually. The marginal percentages for age group were 33.3% in the mature adult category and 66.7% in the young adult category. The conditional distribution of model dress for age shows that not sexual has a higher percentage than sexual for both age groups. A bar chart of this conditional distribution is shown below.

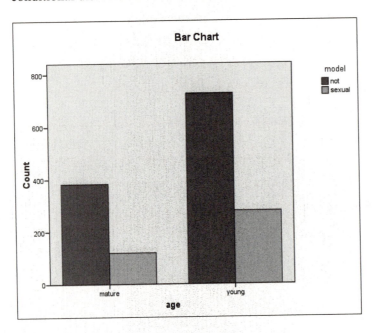

9.39 **a)** $\chi^2 = 76.675$, *P*-value is close to 0 from SPSS. Reject the null hypothesis. There is very strong evidence that there is an association between collegiate sports division and report of cheating. **b)** Answers will vary, but here is one possibility: If we change all the sample sizes to 1000 but keep the percentages the same, $\chi^2 = 15.713$ and the *P*-value remains very close to 0. If we change all the "yes" answer counts to 100 but keep the percentages the same, then $\chi^2 = 7.749$ and the *P*-value increases to 0.021. **c)** The people most likely not to respond are those who gamble, so the results may be biased toward "no" answers. **d)** If one member of a team is cheating, it may be more likely that others are cheating, too. The teammates also may have had a discussion about how to fill out the form.

9.41 $\chi^2 = 12$, *P*-value = 0.001 from SPSS. Reject the null hypothesis. There is strong evidence to say that there is an association between gender and visits to the *H. bihai* flowers. The 0 cell count does not invalidate the significance test. It is the expected counts that need to be 5 or greater for a 2×2 table in order for the chi-square test to be appropriate.

Gender	Visits *H. bihai*		Total
	Yes	No	
Female	20	29	49
Male	0	21	21
Total	20	50	70

9.43 **a)**

	Yes	No
First-year Students	79,317	105,140
Seniors	97,429	97,429

b) H_0: There is no association between grade level of students and whether or not they think statistical data needs to be explained; H_a: There is an association between grade level of students and whether or not they think statistical data needs to be explained. **c)** $\chi^2 = 1865.752$, df = 1, P-value = 0. There is evidence that there is an association between the class of a student and whether or not they think that an explanation is required for statistical or numerical information. **d)** The data do show that seniors are more likely to believe that an explanation is required for statistical or numerical data. However, instructors might wish for a greater change in the percent who believe this than what is seen between these two groups of students.

9.45 $\chi^2 = 50.81$, df = 2, $P < .0005$. The older employees appear to receive lower performance evaluations. They are twice as likely to fall into the lower performance category but only 1/3 as likely to fall in the highest category.

9.47 For Exercise 9.10—data from a single sample; Exercise 9.12—data from a single sample; Exercise 9.16—comparison of several populations based on separate samples; Exercise 9.28—data from a single sample; Exercise 9.30—data from a single sample.

9.49 **a)**

	NoRDC	YesRDC
Asset Size Under 100	257.5	114.5
Asset Size 101 to 200	132.2	58.8
Asset Size 201 or More	136.3	60.7

b) $\chi^2 = 96.305$. **c)** df = 2. **d)** P-value < 0.0005. The largest value on Table F for 2 degrees of freedom is 15.20. The associated P-value is given as P-value < 0.0005.

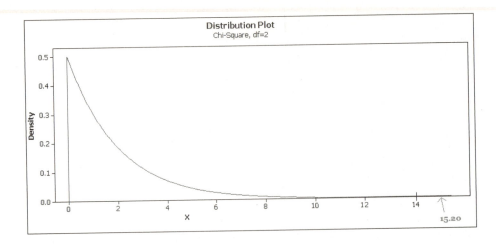

Case Study 9.1

Answers will vary depending on selection of variables that interest the student.

Case Study 9.2

Employment Status
The following is a table of counts for satisfaction of students who are employed/not employed by field of study. Note that numbers in total column do not match text exactly due to rounding.

Field	Employed	Not Employed	Total
Trades	236	707	943
Design	270	329	599
Health	2,198	3,036	5,234
Media/IT	1,036	2,202	3,238
Service	524	854	1,378
Other	874	1,426	2,300
Total	5,138	8,554	13,692

Overall, 62.47% of Canadian students in private career colleges are employed. This does vary significantly by fields as seen by the chi-square test where $\chi^2 = 164.606$, df = 5, and the *P*-value ≈ 0. The following graph shows the number employed and the number not employed for each field. Notice that there are a greater number unemployed in each field than employed but that the difference between the two categories changes by field of study. For design, the percent unemployed and the percent employed are much closer together than the same percents for the "trades" field. Health has the greatest number of employed students and the greatest number of unemployed students, while trades has the smallest number of employed students and design has the smallest number of unemployed students.

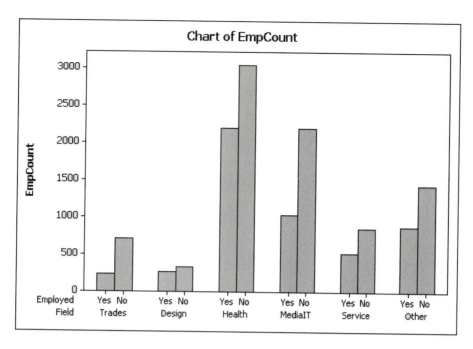

Salary Range

The following is a table of counts for salary ranges of those students employed full time.

Field	$19 or Less	$20 to $29	$30 to $39	$40 or More	No Response	Total
Trades	31	12	4	17	5	69
Design	13	4	1	1	1	20
Health	153	76	38	35	19	321
Media/IT	133	76	28	34	11	282
Service	50	23	15	10	5	103
Other	50	25	16	10	4	105
Total	430	216	102	107	45	900

A chi-square test of the association between the field of study and the salary range yields $\chi^2 = 23.068$, df = 20, and the *P*-value = 0.285. There is not evidence of an association between these two variables. Therefore, the marginal distributions should be sufficient to discuss these results. The marginal distribution of salary range is 47.8% for $19 or less, 24.0% for $20 to $29, 11.3% for $30 to $39, 11.9% for $40 or more, and 5.0% nonresponse. When the nonresponse results are removed, we still get a non-significant result from the chi-square test, but the test statistic is smaller at 20.995 with a *P*-value of 0.137. The graph that follows shows the distributions of salary across the fields of study with the non-response category included.

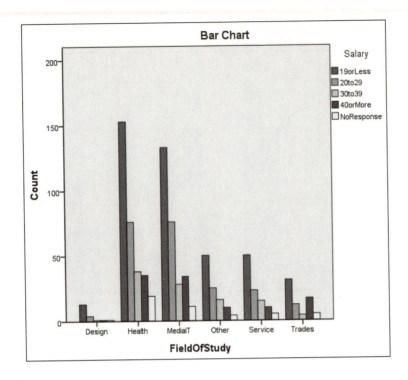

Preprogram Employment Status

The following is a table of counts for the preprogram employment status of Canadian students in private career colleges. Note that numbers in total column do not match text exactly due to rounding.

Field	Employed	Not Working	In School	Other	No Response	Total
Trades	405	217	104	47	170	943
Design	294	84	108	36	72	594
Health	2,565	680	628	471	890	5,234
Media/IT	1,263	777	356	356	486	3,238
Service	634	138	110	152	262	1,296
Other	1,150	483	207	138	276	2,254
Total	6,324	2,386	1,516	1,202	2,156	13,559

The results of a chi-square test for this data show that $\chi^2 = 437.274$, df = 20, and the *P*-value ≈ 0. There is evidence of an association between the field of study and the preprogram employment status. The marginal distribution for preprogram employment status is 46.5% employed, 17.5% not working, 11.2% in school, 8.9% other, and 15.9% no response. Students in the design field were more likely to have been in school (18.8%) than any of the other fields. Those in the service field were least likely to have been in school (8.5%). Media/IT students were in the field least likely to have been employed prior to the program (39.0%) and most likely to be "not working" (24.0%). Those in the service field were more likely to answer "other" than the remaining fields at 11.7%. This field also had the highest nonresponse rate. The graph that follows shows the counts for the various preprogram employment options by field of study.

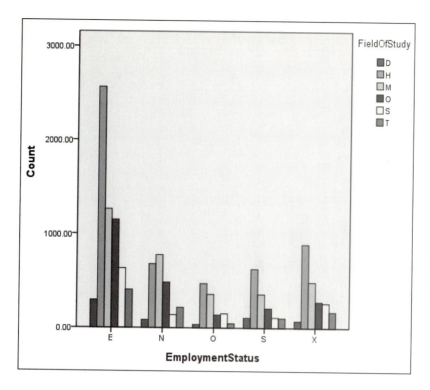

Satisfaction with Course Content

The following is a table of counts for satisfaction of students with job market preparation.

Field	Satisfied	Not Satisfied	Total
Trades	782	160	942
Design	515	84	599
Health	4,449	785	5,234
Media/IT	2,688	550	3,238
Service	1,199	41	1,378
Other	2,070	230	2,300
Total	11,703	1,988	13,691

The chi-square test shows a strong indication of an association between field and satisfaction with course content: $\chi^2 = 62.014$, df = 5, and the *P*-value ≈ 0. In general, 85.5% of students are satisfied with the course content, but this is as low as 83% for media and for trades and as high as 90% for other. Following is a graph of the count who are satisfied and those who are not satisfied with the course content by field. Notice particularly that the graph does not show a large difference in the number who are not satisfied with the course content for other, service, and trades, but the graph does show a large difference for these same fields in the number who are satisfied with the content.

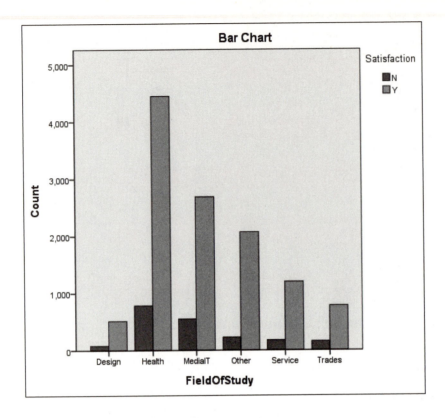

Satisfaction with Job Market Preparation
The following is a table of counts for satisfaction of students with the course content.

Field	Satisfied	Not Satisfied	Total
Trades	716	226	942
Design	449	150	599
Health	4,030	1,204	5,234
Media/IT	2,396	842	3,238
Service	1,089	289	1,378
Other	1,748	552	2,300
Total	716	226	13,691

Overall, 23.8% of Canadian students were not satisfied with their job market preparation while 76.2% were. The chi-square test shows that there is an association between whether a student was satisfied and their field of study. The test statistic was $\chi^2 =$ 17.131, df = 5, and the P-value = 0.004. Those in the fields of media and service contributed the most to the chi-square statistic. Media had fewer students who were satisfied with their preparation, while service had more who were satisfied than the general population. The graph that follows shows the distribution of those who were satisfied and those who were not satisfied with their job market preparation across the fields of study.

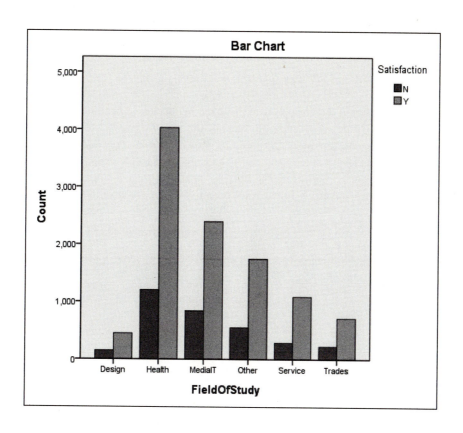

Chapter 10: Inference for Regression

10.1 Predicted stress level is 2.3752. The residual is 0.3048.

10.3 **a)** $\beta_0 = 0.3$. The average overseas return will be 0.3 when the U.S. market is flat.
b) $\beta_1 = 0.85$. There will be an average increase of 0.85% in overseas returns for each 1% increase in return on common stocks in the U.S. **c)** *Mean Overseas Return* = 0.3 + 0.85 × *U.S. Return* + ε. ε allows the overseas returns to vary for a fixed value of U.S. returns.

10.5 **a)** Yes, there is a strong, positive, linear relationship between year and spending.

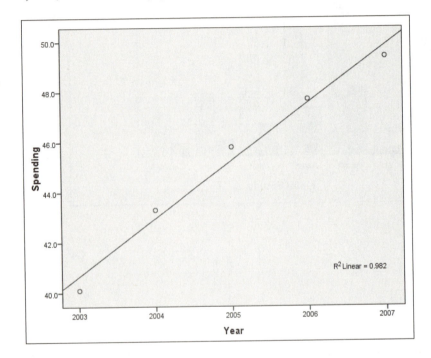

b) $\hat{y} = -4566.24 + 2.3x$. **c)** Residuals are shown in the table below.

2003	2004	2005	2006	2007
−0.56	0.34	0.54	0.14	−0.46

d) *SPENDING* = $\beta_0 + \beta_1 \cdot YEAR$ + ε, with estimates $\beta_1 = 2.3$, $\beta_0 = -4566.24$, and ε = 0.563.
e) For 2001, the predicted spending is $36.06. The residual is $3.26. This is an extrapolation. The trend might not have been the same prior to 2003.

10.7 $b_1 = 0.6269$, $SE_{b1} = 0.0889$, $H_0: \beta_1 = 0$, $H_a: \beta_1 > 0$, $t = 7.052$, df = 49, *P*-value ≈ 0. There is significant evidence to conclude that β_1 is greater than 0.

10.9 Chance plays a role in the performance of a fund. The group of mutual funds that performed well last year will be influenced by chance this year and will likely see a smaller return than last year.

10.11 **a)** $r = 0.7096$, $t = 7.05$, df $= 49$, P-value < 0.0005. Based on this P-value, it is reasonable to conclude that there is a positive correlation between return on treasury bills and inflation. **b)** Verify.

10.13 **a)** 22 have selling prices greater than the assessed values. Not all markets will see this trend. In 2008 to 2010, for example, the selling prices of homes decreased quite a bit compared to their assessed values. **b)** The scatterplot shows a strong, positive, linear relationship between the variables as seen below.

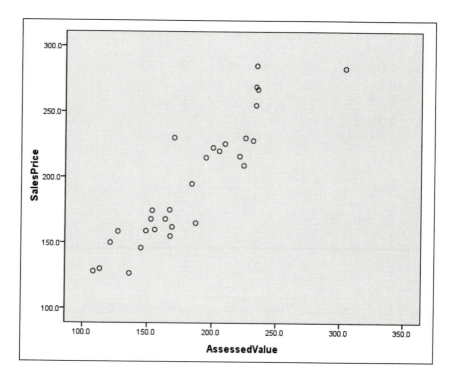

c) $\hat{y} = 21.499 + 0.947x$. **d)** The residual plot below shows no indication that there is a problem with linearity in the data.

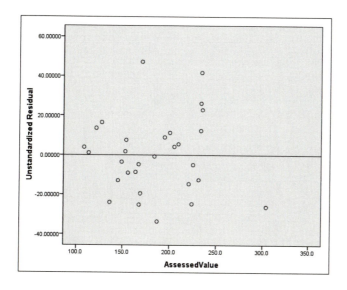

e) A Normal probability plot shows that the points follow a 45-degree angle line, indicating Normality of the residuals. **f)** Both the linearity assumption and the Normality of the residuals assumption are met.

10.15   **a)** The scatterplot below shows a positive linear relationship between the tuitions for these two years. No outliers are apparent in the graph.

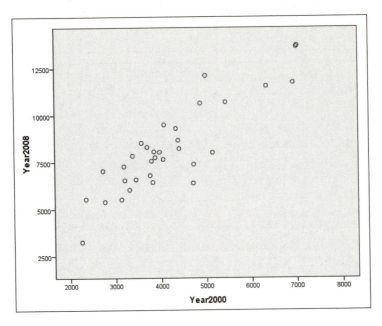

b) $\hat{y} = 1132.75 + 1.692x$. **c)** The residual plot shown below shows random scatter of the residuals.

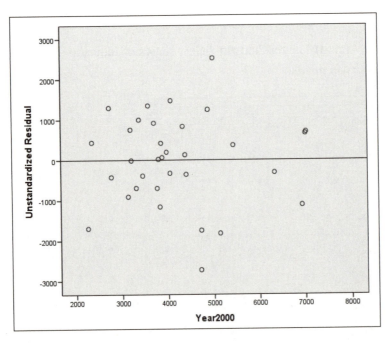

d) The Normal probability plot does show a pattern with occasional jumps in the graph followed by three to six points where the cumulative probability does not increase at the rate expected and then another jump. This would not be expected in a Normal distribution.

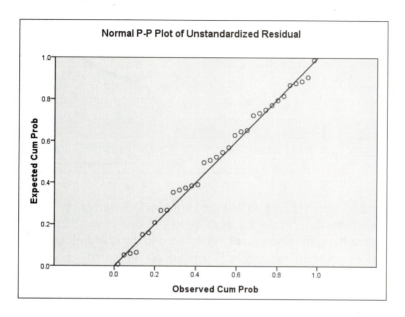

e) H_0: $\beta_1 = 0$, H_a: $\beta_1 \neq 0$. **f)** The test statistic is $t = 10.552$ with a P-value = 0. There is evidence that there is a linear relationship between the 2000 and 2008 tuitions.

10.17 **a)** As seen in the graphs below, the percentage of the salary from incentive payments is strongly skewed to the right and non-Normal. The median is 1.4286 and the mean is 14.2205. The range of values is from 0 to 85.015.

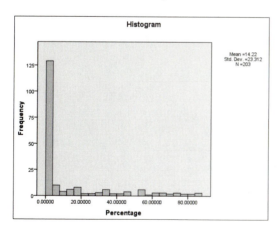

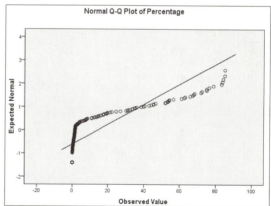

The graphs below show that the rating is also skewed to the right, but less strongly skewed than the percentage of salary from incentive payments. The distribution is also not Normal. The median rating is 6.31, the average is 7.759, and the range is from 0 to 27.88.

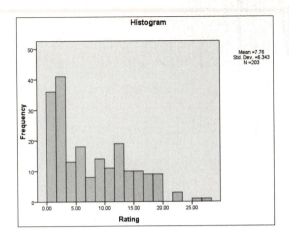
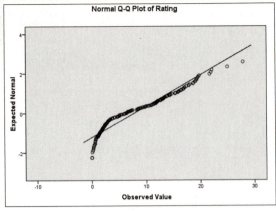

b) The regression model does not require that the variables themselves be Normally distributed. It requires that the *y* values for any fixed value of *x* be Normally distributed with the same standard deviation throughout. **c)** As seen in the scatterplot that follows, a linear relationship does not seem reasonable. There are no particularly unusual values, but the relationship is very weak.

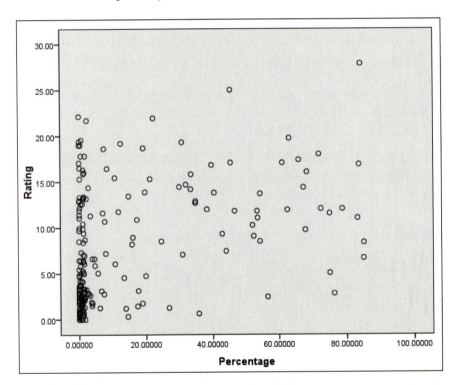

d) The least-squares regression line is $\hat{y} = 6.247 + 0.106x$. **e)** The Normal probability plot below shows that the distribution of the residuals is not Normal.

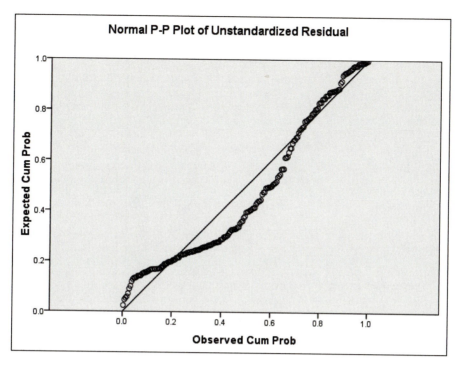

Normal P-P Plot of Unstandardized Residual

10.19   a)

Variable	Area	IBI
Mean	28.29	65.94
S	17.714	18.280
Min	2	29
Max	70	91
	Fairly symmetric and Normally distributed except for 2 high outliers	Skewed left but no outliers

b) The scatterplot shows a relationship that is positive and weakly linear. It's hard to tell if there are any outliers or unusual patterns because the relationship is so weak.

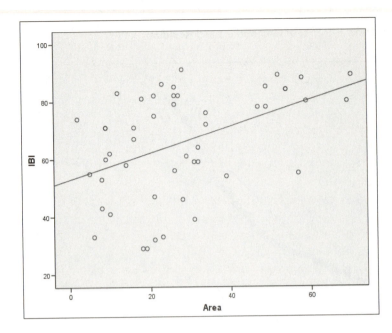

c) $IBI = \beta_0 + \beta_1 \cdot AREA + \varepsilon$. **d)** $H_0 : \beta_1 = 0$, $H_a : \beta_1 \neq 0$. **e)** $\hat{y} = 52.923 + 0.460x$.
From SPSS, the hypothesis test in part (d) has $t = 3.415$ and a P-value of 0.001, so we can reject the null hypothesis. There is strong evidence that Area and IBI have a linear relationship. $r^2 = 19.9\%$, and the regression standard error is $s = 16.535$. **f)** The residual plot shows that the residuals get slightly less spread out as Area increases, but overall, it doesn't look too bad. **g)** Yes, the Normal probability plot looks good.

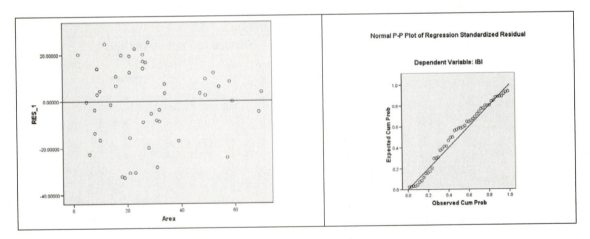

h) The assumptions are reasonable—we are assuming a simple random sample from the population, there is a fairly linear (although weak) relationship between IBI and area, there is approximately the same spread above and below the regression line on the scatterplot, which is fairly uniform for all area values on the plot, and the Normal probability and residual plots look good.

10.21 Area is the better explanatory variable for regression with IBI. IBI and forest have a much weaker relationship than IBI and area.

10.23 $H_0 : \rho = 0$, $H_a : \rho \neq 0$, P-value $= 0.061$, so do not reject the null hypothesis. There is not enough evidence to say that the correlation is significantly different from 0 if a significance level of 0.05 is used. This agrees with what we saw in the test using the slope in Exercise 10.20. The correlation is a numerical description of the linear relationship between forest and IBI, which is very weak.

10.25 **a)** The vertical stacks appear because age has been truncated to the nearest year. This results in many x's with the same numerical values. **b)** Older men have more experience. Younger men may have a more current education. It does not appear that age has a strong relationship with income. **c)** $\hat{y} = 24874 + 892x$. The slope tells us that, for every year a man ages, his income increases by \$892 on average.

10.27 **a)** The close cluster of values between 0 and 100,000 and the scattered points above 100,000 indicate a distribution skewed to the right for each value of x. **b)** Large sample sizes help make up for any skewness in the sample data.

10.29 **a)** The intercept represents the return on Treasury bills when there is no inflation. We would expect a positive return on any money invested. **b)** $b_0 = 2.9645$. The standard error is 0.4452. **c)** Both the lower and upper bounds for the intercept are positive, so there is good evidence that β_0 is greater than 0. **d)** $2.9645 \pm 2.021 \times 0.4452$, using degrees of freedom of 40 to be conservative.

10.31 A correlation of –0.36 means that when establishments close to campus charge a higher rate for a bottle of beer, the binge-drinking rate will be less on average than when the establishments charge a lower price. Since the correlation is closer to zero than to 1, this is not a very strong relationship, so this relationship will not hold true in every case.

10.33 **a)** There is a moderate, positive, linear relationship between size and price. The r^2 is 43.1%. Yes, size is helpful in predicting selling price.

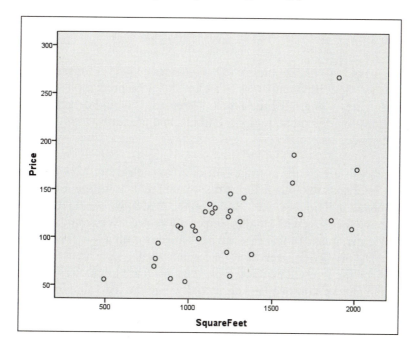

b) $\hat{y} = 21.398 + 0.077x$, $t = 4.601$, P-value < 0.0001. The small P-value means that there is a significant linear relationship between size and selling price.

10.35 $r = 0.656$, $t = 4.601$, P-value < 0.0001. Conclude that the population correlation is greater than zero, and therefore larger houses have higher prices.

10.37 Removing the point (1897, 268) results in an r^2 of 38.0%, $t = 4.068$ with a very small P-value, and the equation of the line is $\hat{y} = 40.647 + 0.058x$. While the equation of the line changed, the conclusions are the same as they were before, so this outlier is not extremely influential.

10.39 **a)**   **b)**

c) $Log\hat{B}its = -872.928 + 0.446\ year$. (0.432, 0.461).

10.41 **a)** Verify. **b)** 2.894 ± 0.120.

10.43 **a)** (87.87%, 108.60%). **b)** (8543, 13,354). **c)** (13,085, 18,628). **d)** The prediction interval is wider for Moneypit U because we have less information at this extreme of a value for the 2000 tuition.

10.45 **a)** (69.34, 86.21). **b)** (43.46, 112.09). **c)** A 95% confidence interval for mean response is the interval for the average IBI for all areas of 54 km^2. A 95% prediction interval for a future response estimates a single IBI for a 54 km^2 area. **d)** It would depend on the type of terrain and the location. Mountain regions, latitude, and proximity to industrial areas might make a difference.

10.47 The confidence intervals are for the average IBI for areas with either 22% forest or areas of 54 km^2. These intervals are quite different due to the fact that we are looking for an "average" value, not an individual value. Based on the model, we would expect to see lower IBIs on average for areas with smaller percents forested. We would also expect to see higher IBIs in larger areas. In this case we have an area with a low forested percent and a larger area. The prediction interval instead is intended to give an estimate for a future IBI value given the forested percent or the area in square kilometers. We see here that these prediction intervals do overlap substantially, and the IBI value for a similar area would likely be in the overlapping portion of these two intervals.

10.49 **a)** Verify. **b)** (49,780, 53,496). **c)** (−41,735, 145,010). This interval is not very useful.

10.51 (58,918, 62,200).

10.53 **a)** (44,446, 55,314). **b)** More data was used to develop the confidence interval in exercise 10.49 than was used in part (a) of this problem. Since standard errors are a function of sample size, when we have larger samples we will have smaller standard errors.

10.55 The 90% prediction interval for Steve's BAC is (0.04, 0.11424). He should not drive.

10.57 $SE_{b1} = 0.0889$.

10.59 $H_0: \beta_1 = 0$; $H_a: \beta_1 \neq 0$, $t = 7.0501$, $F = 49.705$, $t^2 = F$, P-value = 5.55E-09.

10.61 **a)** $r^2 = (174.956/347.431) = 0.5036$. **b)** $s =$ **Error! Bookmark not defined.** $\sqrt{3.520} = 1.8761$.

10.63 $r^2 = (4560.6/8556.0) = 0.533$, $s = \sqrt{221.967} = 14.899$.

10.65 **a)** $H_0: \beta_1 = 0$, $H_a: \beta_1 \neq 0$, $t = 6.041$, P-value = 0.0001. Yes, these data give a statistically significant result. Reputation helps explain profitability. **b)** $r^2 = 0.1936$. 19.36% of the variation in profit can be explained by the reputation of the company. **c)** Statistical significance does not always translate in practical significance. While there is some relationship between reputation and profitability, reputation explains a small percentage of the variation in profitability. There are likely many more variables that would help predict profitability of a company.

10.67 The 95% confidence interval for mean profitability for companies with a reputation of 7 is (0.110604, 0.141804) with appropriate round-off error.

10.69 The F statistic is equal to the t statistic squared. The P-values will be the same. $F = 36.492 = (6.041)^2$.

10.71 $r^2 = (SS_{model}/SS_{total}) = 0.1936$.

10.73 **a)** The slope describes the average change in y for a one unit change in x. **b)** The population regression line is $\mu_y = \beta_0 + \beta_1 x$. **c)** The 95% confidence interval varies in width depending on the specific value of x.

10.75 **a)** Based on the scatterplot below, it does appear that there is a linear relationship between the driver's calculations of mpg and the computer's calculation of mpg.

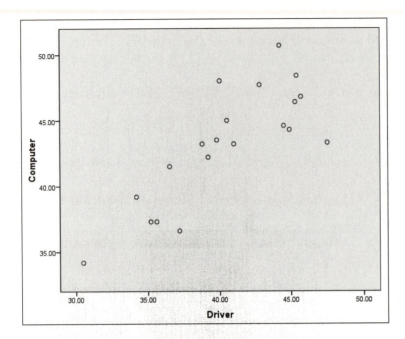

b) $\hat{y} = 11.812 + 0.775x$. **c)** It does not appear that the calculations are the same. If they were, the slope of the regression line would have been close to 1, and the intercept of the regression line would have been close to 0.

10.77   Answers will vary.

10.79   **a)** For more expensive items, the pharmacy would be charging less of a markup.
b) $\hat{y} = 2.885 - 0.295x$, where $\hat{y}$ is the predicted (log) markup and x is the (log) cost.
c) df = 137, but using the t table, we would use df = 100 to be conservative. For testing $H_0 : \beta_1 = 0$, $H_a : \beta_1 < 0$, the P-value is less than 0.005, so we can reject the null hypothesis. There is strong evidence that (log) markup and (log) cost have a negative linear relationship (evidence that charge compression is taking place).

10.81   **a)** Answers will vary. Some questions to consider: Is this a national chain or an independent restaurant? What kind of food is prepared? Do the restaurants have similar staffing and experience? **b)** Answers will vary.

10.83   **a)** The relationship seen in the scatterplot below looks positive, linear, and moderately strong except for potential outliers for the two largest hotels (1388 and 1590 rooms each).

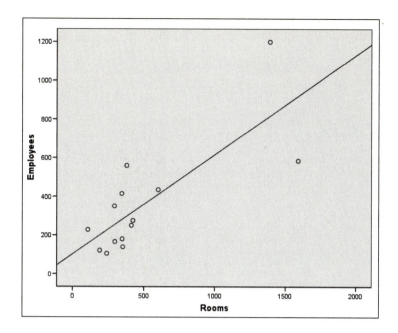

b) The moderately strong linear relationship is good, but the outliers are not.
c) $\hat{y} = 101.981 + 0.514x$, where $\hat{y}$ is the predicted number of employees and x is the number of rooms. **d)** $H_0 : \beta_1 = 0$, $H_a : \beta_1 \neq 0$ with $t = 4.287$ and a P-value from SPSS of 0.001, so reject the null hypothesis. There is strong evidence that there is a significant linear relationship between the number of rooms and the number of employees working at hotels in Toronto. **e)** (0.253, 0.775).

10.85 **a)** Hotel 1 (1388 rooms) and Hotel 11 (1590 rooms) are the two outliers. **b)** The r^2 drops from 60.5% in the original model to just 26.3% in the model without the outliers. The s drops from 188.774 in the original model to 129.151 in the model without the outliers. The new equation of the line is $\hat{y} = 72.526 + 0.589x$. The new test statistic is $t = 1.891$, and the P-value is now 0.088, so we cannot reject the null hypothesis. There is not enough evidence to say that the slope is significantly different from 0. The 95% confidence interval for the slope is (–0.105, 1.284), which contains 0. The 95% confidence interval for the slope from the original model did not contain 0.

10.87 The scatterplot shows a strong, linear, and positive relationship between the number of students and the total yearly expenditures.

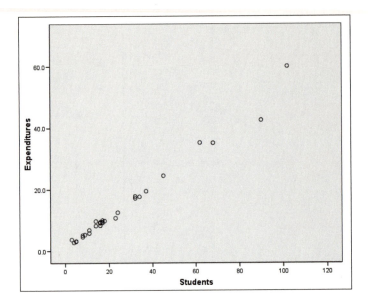

The least-squares regression line is $\hat{y} = 0.526 + 0.530x$, where $\hat{y}$ is the predicted total yearly expenditure and x is the number of students. r^2 is very good at 98.4%, and the regression standard error is $s = 1.70$. A two-sided test of the slope gives a $t = 41.623$ and a P-value close to 0. These results sound very promising; however, the Normal probability plot does not look good, and the residual plot has a distinct funnel shape. The assumptions for regression may not be appropriate here.

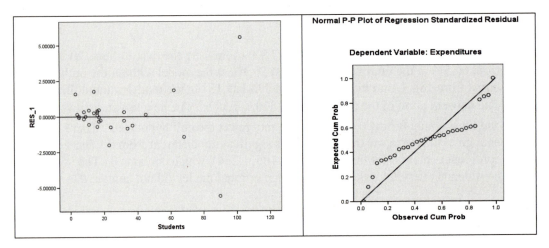

10.89 $H_0: \beta_1 = 0$, $H_a: \beta_1 \neq 0$, $t = 2.16$, $0.02 < P$-value < 0.04. Yes, there is a significant linear relationship between pretest and final exam scores.

Case Study 10.1

a)

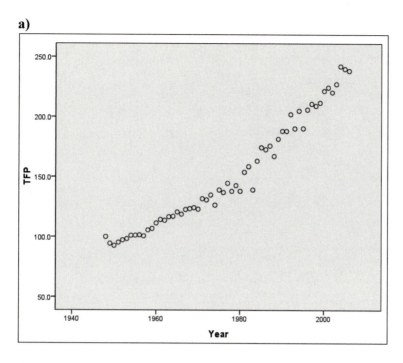

Around 1980, the rate of increase of TFP started to go up. TFP started to increase at a faster rate. The variation also appears to have increased. **b)** $\hat{y} = -3052.511 + 1.614x$.

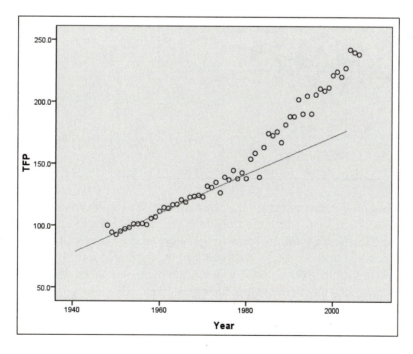

c) (1.495, 1.733). **d)** $\hat{y} = -6943.986 + 3.582x$. (3.218, 3.946).

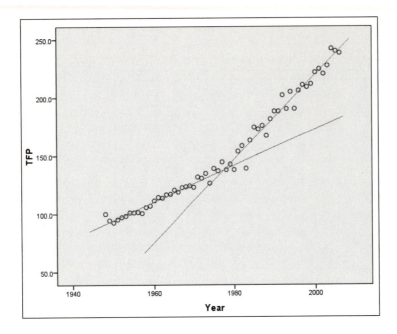

Case Study 10.2

a)

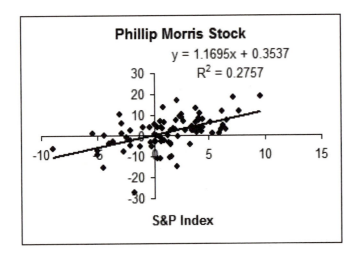

The slope of the regression line indicates that the return on Phillip Morris stock increases when the S&P index increases. Since the slope is greater than 1, this indicates that the returns for Phillip Morris stock will increase (or decrease) more sharply than the market in general. The positive intercept indicates that when the market is flat, we would still expect an increase in Phillip Morris stock. **b)** Each sample that is taken will result in somewhat different results unless the values are repeated exactly. Therefore, we would expect future results to have some variation from what was seen in the initial results, even if there were no changes to the company's general position in relationship to the market. (0.82, 1.52). **c)** Data points 33 and 73 have large negative residuals. These are not likely to be influential because many other data points anchor them above. **d)** Aside from the two low outliers, the residuals appear fairly Normal. **e)** (6.653, 11.597).

Chapter 11: Multiple Regression

11.1 $\hat{y} = 210 + 160(45.25) + 160(15.50) + 150(41.52) + 65(411.88) + 120(25.24) = 45,959 \text{ ft}^2$.

11.3   **a)** The response variable is bank assets. **b)** The explanatory variables are the number of banks and deposits. **c)** There are two explanatory variables; therefore, $p = 2$. **d)** The sample size is 52.

11.5   The variables do not need to have a Normal distribution. The requirement is that the residuals should be Normally distributed for multiple regression.

11.7

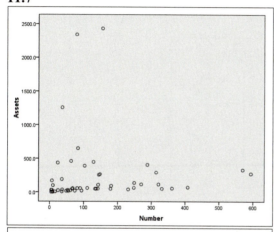

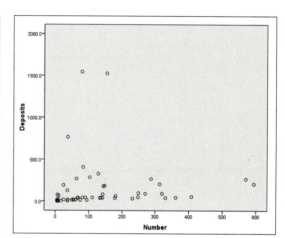

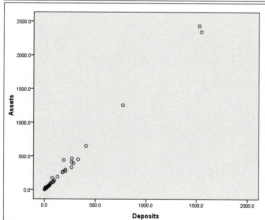

The relationship is definitely strongest between assets and deposits. The correlations are 0.018 between number of banks and assets, 0.998 between deposits and assets, and 0.049 between deposits and the number of banks. The states mentioned in the previous problem stand out as being outliers in these graphs as well.

11.9 $\widehat{\text{Assets}} = 8.090 - 0.108 \times \text{Number} + 1.566 \times \text{Deposits}$

11.11

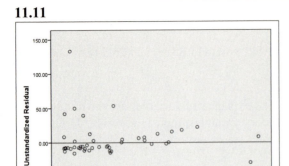

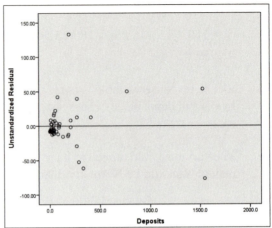

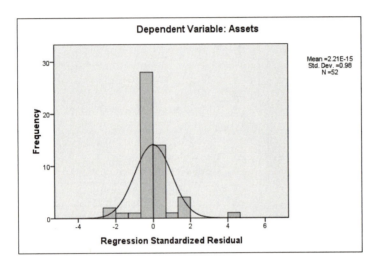

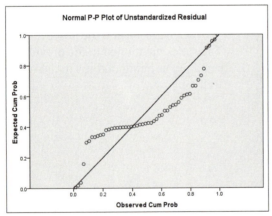

The outliers are highlighted again by the residual plots. The histogram shows that the residuals have a greater portion of their observations in the center of the distribution and fewer observations in the tails than what would be seen in the Normal distribution. The Normal probability plot shows that the distribution of the residuals is not Normal.

11.13

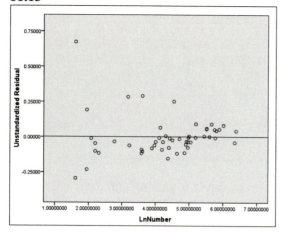

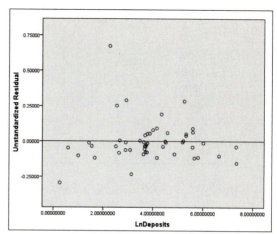

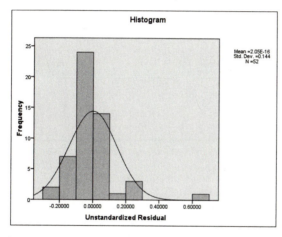

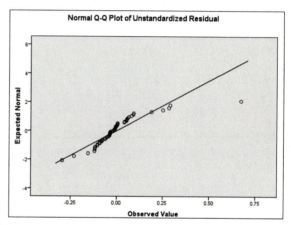

The residual plots show that the data is less clustered at the low values than what was seen in the original data. The data also are closer to the Normal distribution as seen in the histogram and the Normal probability plot. Normality of the residuals is still not met, but it is not as far off as was seen previously.

11.15

	Value of s	Value of s^2	Name of s
Excel	0.354000516	0.12531637	Standard error
SPSS	0.354001	0.125	Std. error of the estimate
SAS	0.35400	0.12532	Root MSE
Minitab	0.354001	0.1253	S

11.17 **a)** The response variable is overall math GPA. **b)** There are 106 cases. **c)** $p = 4$. **d)** The four explanatory variables are SAT math score, SAT verbal score, class rank, and Bryant College's mathematics placement score.

11.19   a)

	Mean	Median	Standard Deviation
Price	116.58	120	47.142
Weight	12.316	12	1.945
Amps	13.711	15	1.7104
Depth	2.4112	2.438	0.0669
Speed	4.053	4	1.026
Power	4.263	4	0.806
Ease	3.632	4	0.597
Construction	4.263	5	0.872

b) None of the distributions shown below have any outliers, although there are a couple of values that are close in the weight distribution. Amps has over half of its distribution at the maximum value. Most of the distributions are skewed to the left. Amps is the only one where this skewness is strong enough to possibly cause concerns.

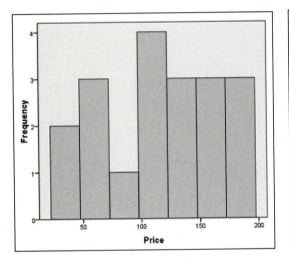

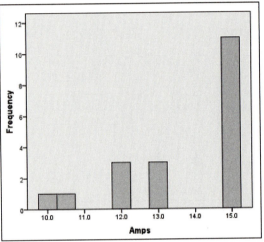

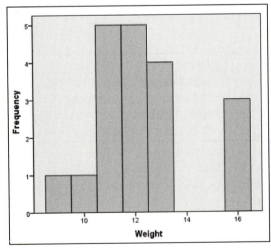

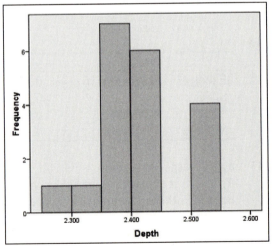

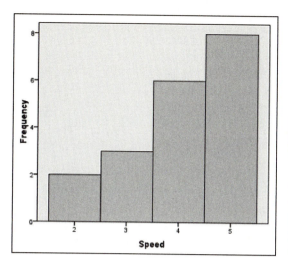

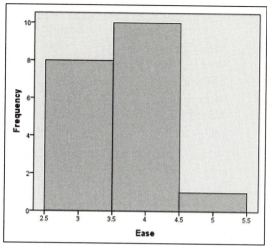

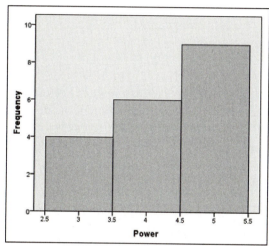

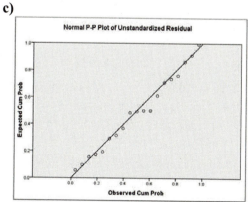

11.21 **a)** $\widehat{Price}$ = -467.715 + 11.567 × Weight + 4.767 × Amps + 82.935 × Depth − 13.275 × Speed − 1.186 × Power + 18.751 × Ease + 39.233 × Construction.
b) $s = 26.642$.
c)

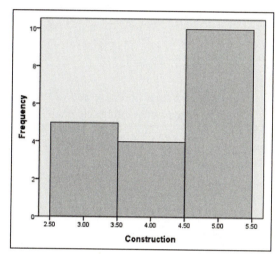

The residuals are not precisely Normal, but there don't appear to be any major problems that would keep us from using the model for these data.

11.23 **a)**

	Accounts	Market Share	Assets
Mean	794	8.96	48.91
Standard Deviation	886	7.74	76.16
Median	509	8.85	15.35
Minimum	125	1.3	1.30
Maximum	2500	27.50	219.00
Q1	134	2.80	5.90
Q3	909	11.60	38.80

b)

Market Share

```
0 | 1  2  3  4
0 | 8  9
1 | 0  2  3
1 |
2 |
2 | 8
```

Accounts

```
0 | 1  1  1  2  4
0 | 6  6  9
1 |
1 |
2 | 3
2 | 5
```

Assets

```
0 | 0  0  0  0  1  2  2  4
0 |
1 |
1 | 6
2 | 2
2 |
```

c) All three stem plots are skewed to the right. The market share data set has one outlier, and both accounts and assets each have two outliers within the distribution of the data set.

11.25 **a)** Market share = $5.159 - 0.000312 \times$ Accounts $+ 0.08277 \times$ Assets. **b)** Regression standard error $s = 5.4876$.

11.27 **a)**

	Market Share	Accounts	Assets
Mean	6.6	392	13.8
Standard Deviation	4.6	292	12.3
Median	6.0	317	9.0
Minimum	1.3	125	1.3
Maximum	12.9	909	38.8
Q1	2.5	132	5.7
Q3	10.8	603	19.8

b)

Market Share

```
0 | 1  2  3  4
0 | 8
1 | 0  2  3
1 |
```

Accounts

```
0 | 1  1  1  2  4
0 | 6  6  9
1 |
1 |
```

Assets

```
0 | 1  6  6  7
1 | 1
2 | 0  1
3 | 9
```

c) The distribution of brokerage firms without the data for Schwab and Fidelity has a much shorter range and no outliers compared to the results of Exercise 11.23. The largest effect was on the mean and standard deviation.

11.29 **a)** Market share = $1.845 + 0.00663 \times$ Accounts $+ 0.1566 \times$ Assets. **b)** Regression standard error $s = 3.50$. The coefficients of the regression equation all changed when a regression analysis was done without Schwab and Fidelity. The standard error *s* was also reduced.

11.31 a)

Gross Total Sales	
Mean	320.30
Standard Error	36.02
Median	263.29
Standard Deviation	180.09
Range	798.20
Minimum	92.30
Maximum	890.50

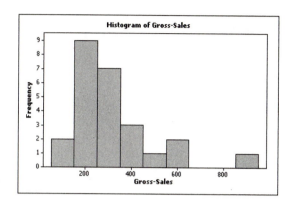

Cash Items	
Mean	20.52
Standard Error	2.36
Median	19.00
Standard Deviation	11.80
Range	50.00
Minimum	5.00
Maximum	55.00

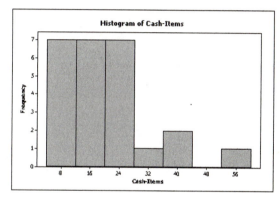

Check Items	
Mean	20.04
Standard Error	2.82
Median	15.00
Standard Deviation	14.07
Range	54.00
Minimum	3.00
Maximum	57.00

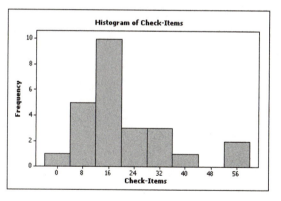

Credit Card Items	
Mean	7.68
Standard Error	1.60
Median	5.00
Standard Deviation	7.98
Range	28.00
Minimum	0
Maximum	28.00

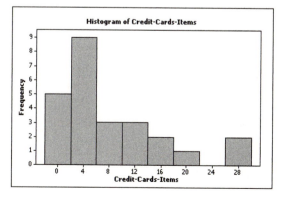

Each of the distributions of the variables is skewed to the right. Both gross total sales and cash items appear to have potential outliers. The distribution for check items has two obvious outliers.

b)

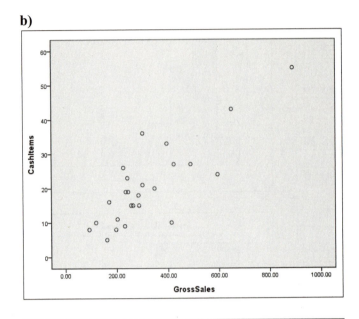

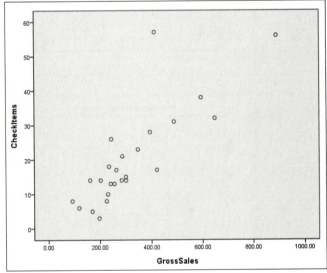

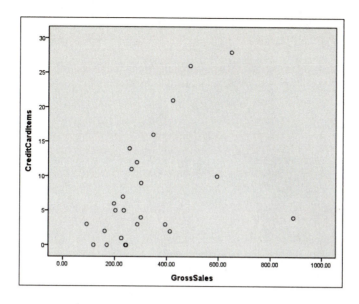

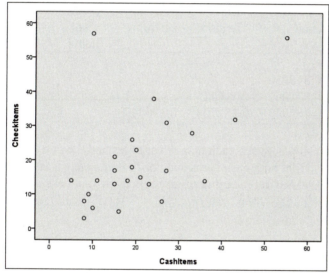

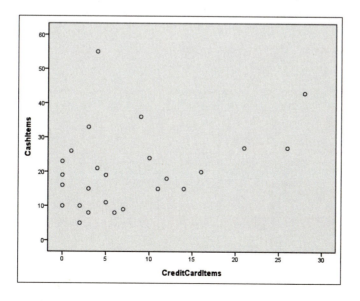

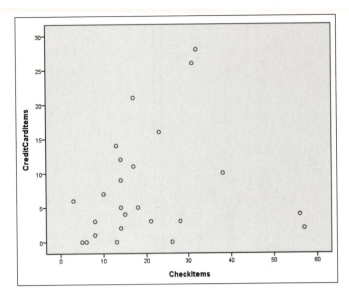

	Gross Total Sales	Cash Items	Check Items	Credit Card Items
Gross Total Sales	1			
Cash Items	0.816956	1		
Check Items	0.821062	0.516425	1	
Credit Card Items	0.457943	0.352708	0.176447	1

After observing the relationship between each pair of variables and the correlation analysis, it can be concluded that both cash items and check items have a fairly strong correlation with gross total sales. Credit card items appears to have a weak correlation with all the other variables. **c)** Gross total sales = 0.341 + 7.100 × Cash items + 6.987 × Check items + 4.458 × Credit card items.

d)

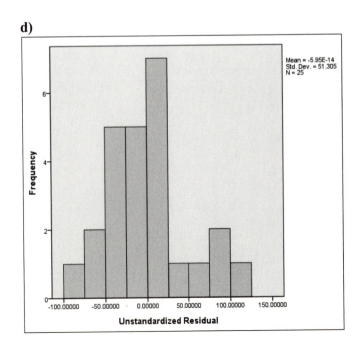

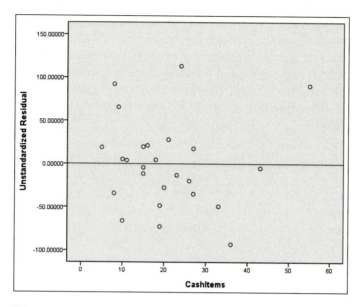

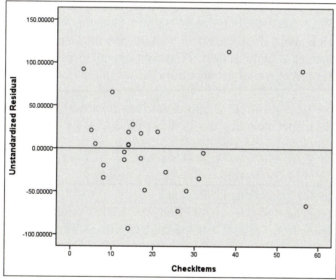

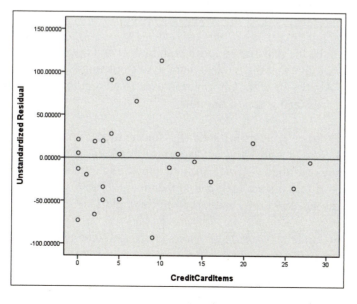

There are no obvious problems with the distribution of the residuals or the plots against the three explanatory variables. **e)** Results from regression analysis are only to be used to predict values for the response variable when the values of the explanatory variables are within the range of those seen in the data used for analysis. The smallest number of items sold in the sample was 16 items in one day. Using the model to predict the gross sales when zero items are sold is extrapolating and will likely result in some error. In this case, the error is small at 34 cents.

11.33

	Excel	SPSS	Minitab
Regression Coefficient for Opening	3.2417285	3.242	3.2417
Standard Error for Opening	0.296467651	0.296	0.2965
t-stat for Opening	10.93451	10.935	10.93
DF	39	39	39
P-value	5.42E-13	0.000	0.000

Opening weekend revenue is a significant variable for predicting revenue.

11.35   Predicted USRevenue = $50.071 - 0.499 \times$ Budget + $3.230 \times$ Opening. With this model, Budget became only slightly more significant than with the previous model. The coefficient of opening changed slightly as well. However, overall there was very little change in the model with the removal of theaters from the list of factors.

11.37   **a)** (76.62, 253.51). **b)** (79.78, 251.34). **c)** The models give similar results; however, the confidence interval for the model including the number of theaters is wider by about 4 million dollars in revenue.

11.39

Variables	Model for Predicteing USRevenue	R^2
Budget, Opening	USRevenue = $50.071 - 0.499$ Budget + 3.230 Opening	0.872
Budget, Theaters	USRevenue = $-360.505 + 0.433$ Budget + 0.134 Theaters	0.446
Opening, Theaters	USRevenue = $21.602 - 0.003$ Theaters + 2.932 Opening	0.850
Budget	USRevenue = $18.792 + 1.136$ Budget	0.169
Opening	USRevenue = $12.268 + 2.901$ Opening	0.850
Theaters	USRevenue = $-368.389 + 0.149$ Theaters	0.426

The best model appears to be the one that includes both budget and opening. This model has the highest R^2 value. Notice that the models that include opening-weekend revenue all have high R^2 values. Almost all of the information used to predict total revenue is contained in the opening-weekend revenue amounts.

11.41 R_1^2 for the model with all three variables is 0.872. R_2^2 for the model with opening removed is 0.446. $F = 119.8125$ with 1 and 36 degrees of freedom. *P*-value ≈ 0. Opening-weekend revenue contributes significantly to explaining total revenue, even when budget and number of theaters are included in the model. The F statistic is approximately equal to the square of the t statistic from Example 11.13.

11.43   In all four cases, $t = 1.986$, and the two-sided *P*-value is between 0.05 and 0.10.

11.45   **a)** The squared multiple correlation, R^2, gives the proportion of the variation in the response variable that is explained by the explanatory variables. **b)** The null hypothesis should be $H_0 : \beta_2 = 0$. **c)** With a small P-value for the ANOVA F test, we would conclude that at least one explanatory variable is significantly different from zero.

11.47   **a)** MSR = 16/3 = 5.3333. MSE = 120/23 = 5.2174. F = MSR / MSE = 1.022. **b)** R^2 = SSR / SST = 16 / 136 = 0.1176.

11.49   **a)** The two movies are *Finding Nemo* with a residual of $109.49 million and *Lord of the Rings: The Return of the King* with a residual of $139.50 million.

b)

	Coefficient	Std Error	*t* statistic	*P*-value
Budget	–0.311	0.139	-2.230	0.032
Opening	2.896	0.208	13.942	0.000
Theaters	0.005	0.013	0.346	0.731

c) There is no change in the significance of the variables, but the coefficients are different from the previous model. Opening has a smaller coefficient and budget's coefficient is not as negative. Theaters now has a positive coefficient instead of a negative one. **d)** As seen below, the residuals are not precisely Normal. Particularly, we note the departure from the Normal distribution of the negative residuals.

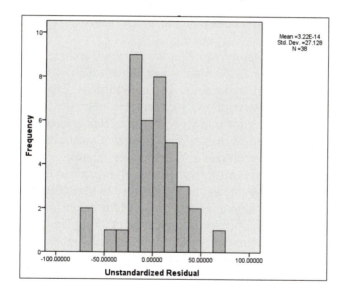

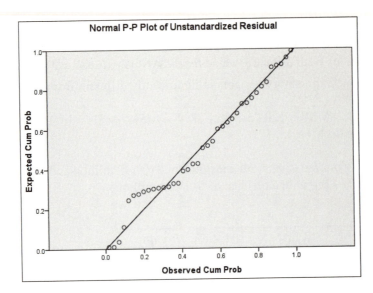

11.51 **a)** $t = -8.917$, df $= 40 - 30 - 1 = 9$, P-value < 0.001. There is strong evidence that the coefficient of P5 is not 0. **b)** $(-1.487, -0.885)$. Since we rejected the null hypothesis in part (a), we did not expect the confidence interval to include 0.

11.53 The difference in signs makes sense because employees generally sign up for either the DC or the DB plan. The total number of dual-earner couples available to sign up for a plan probably stays about the same from year to year; however, as more dual-earner couples sign up for DB plans, fewer would be signing up for DC plans.

11.55 **a)** The hypotheses for each of the explanatory variables are H_0: $\beta = 0$ and H_a: $\beta \neq 0$. The degrees of freedom for the t statistics are 2215. Values that are less than -1.96 or greater than 1.96 will lead to rejection of the null hypothesis. **b)** The significant explanatory variables are loan size, length of loan, percent down payment, cosigner, unsecured loan, total income, bad credit report, young borrower, own home, and years at current address. If an explanatory variable is concluded to be insignificant, then that means the variable does not contribute significantly to the prediction of the response variable. **c)** After examining the signs of each of the 13 explanatory variables with regards to the nature of the variable, it is obvious that a favorable interest rate is awarded to variables that demonstrate some form of lower risk. The interest rate is lower for larger loans, lower for longer length loans, lower for a higher percent down payment; cosigner, lower when there is a cosigner, higher for an unsecured loan, lower for those with higher total income, higher when there is a bad credit report, higher when there is a young borrower, lower when the borrower owns a home and lower when the years at current address is higher.

11.57 **a)** The hypotheses about the jth explanatory variable are H_0: $\beta_j = 0$ and H_a: $\beta_j \neq 0$. The degrees of freedom for the t statistics are 5650. At the 5% level, values of that are less than -1.96 or greater than 1.96 will lead to rejection of the null hypothesis. **b)** The statistically significant explanatory variables are loan size, length of loan, percent down payment, and unsecured loan. **c)** The interest rate is lower for larger loans, lower for longer length loans, lower for a higher percent down payment, and higher for an unsecured loan. Again, these results indicate banks' tendency to give lower interest rates to loans that demonstrate lower risk of default. For example, an unsecured loan is more risky than a secured loan.

11.59 **a)** The *F* statistic has 8 and 494 degrees of freedom. **b)** 68%. **c)** $F = 20.58$ with a *P*-value < 0.0001. There is evidence that at least one of the other variables is helpful in predicting satisfaction when quality of service received and the possibility of negotiating the terms of financing are included in the model.

11.61 **Price** (leaf unit $1000)

```
     5  22
     6  2459
     7  2233566
     8  01124444779999
     9  234469
    10  4
    11  449
    12  4499
    13
    14
    15
    16
    17  39
    18
    19  9
```

Sq Ft (leaf unit 100sf)

```
     0  67777
     0  88899999
     1  000011
     1  22223333
     1  455555
     1  66666
     1  89
     2  01
     2  22
```

The seven homes excluded do appear to skew the distributions to the right.

11.63 **a)** Price $= 45,298 + 34.32 \times$ SqFt. For SqFt $= 1000$, Predicted price $= \$79,622$. For SqFt $= 1500$, Predicted price $= \$96,783$.

11.65 Price $= 81,273.37 - 30.14 \times$ SqFt $+ 0.027 \times$ SqFt2. For SqFt $= 1000$, Predicted price $= \$78,253$. For SqFt $= 1500$, Predicted price $= \$97,042$. Overall, the two sets of predictions are fairly similar.

11.67 Minitab, $t = -2.88$, *P*-value $= 0.0068$, df $= 35$. These values agree with Example 11.21.

11.69

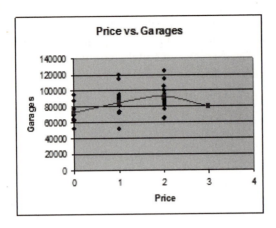

The trend is increasing and roughly linear as one examines from 0 to 2 garages and then levels off.

11.71 When a 1400 ft^2 home has an extra half bath, the average price is \$99,719. When the same size home does not have an extra half bath, the average price is \$84,585. The difference in price is \$15,134.

11.73 No, it does not make sense to compare homes with whole and half baths less than 700 ft^2 because in this study no homes that small have a half bath.

11.75

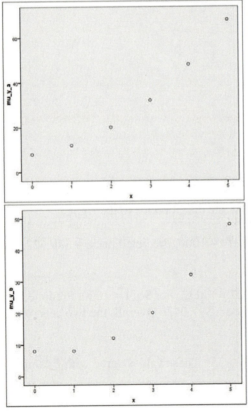

a) The relationship is increasing and concave up.

b) The relationship is increasing and concave up, but the curve is not as steep as in part (a).

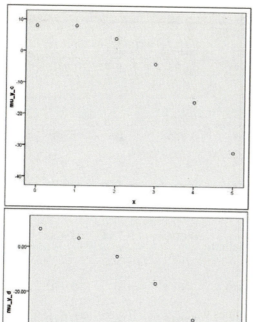

c) The relationship is decreasing and concave down.

d) The relationship is decreasing and concave down, but the curve is steeper than the one in part (b).

11.77 For each equation, the difference in means is equal to the coefficient of x. This will be true in general.

11.79 For part (a), the difference in slopes is $12 - 6 = 6$. This is the coefficient of x_1x_2. The difference in intercepts is $100 - 70 = 30$, which is the coefficient of x_1. Similar calculations can be made for parts (b) and (c). This is true in general.

11.81 **a)** Assets $= 7.6 - 0.00457 \times$ Account $+ 3.36 \times 10^{-5} \times$ Account^2. **b)** $(1.25 \times 10^{-5}, 5.47 \times 10^{-5})$. **c)** $t = 3.76$, df $= 7$, P-value $= 0.007$. The squared term lends predictive power to the model. **d)** Since the variables account and account^2 are dependent (or correlated), the model without the squared term will give a different coefficient on the account variable.

11.83 **a)** $F(2, 37) = 16.223$ with a P-value of 0, so at least one of the variables is significant in the model that includes both theaters and the square of theaters. However, the test statistics for the individual variables (P-values of 0.228 for theaters and 0.099 for theaters^2) are not significant at the 5% level, indicating that the combination is significant, but not the individual variables. This result looks strange and is due to the dependence of the variables. **b)** R^2 from the model with both theaters and the square of theaters is 0.467. R^2 from the model with only theaters is 0.426. $F = 2.846$. The P-value is 0.100, indicating that theaters does not contribute significantly to explaining U.S. box office revenue when theaters^2 is already in the model. **c)** Verify.

11.85

Variables	COEFFICIENTS					R^2	s
	Constant	Opening	Budget	Theaters	Sequel		
Opening, Budget, Theaters, Sequel	23.951	2.856 (*)	−0.317 (*)	0.004	5.207	0.930	28.673
Opening, Budget, Theaters	22.050	2.896 (*)	−0.311 (*)	0.005	0	0.930	28.299
Opening, Budget, Sequel	36.691	2.896 (*)	−0.317 (*)	0	5.523	0.930	28.291
Opening, Theaters, Sequel	0.811	2.681 (*)	0	0.004	0.758	0.920	30.298
Opening, Budget	35.641	2.942 (*)	−0.310 (*)	0	0	0.930	27.941
Opening, Theaters	0.600	2.688 (*)	0	0.004	0	0.920	29.863
Opening, Sequel	12.289	2.717 (*)	0	0	1.051	0.919	29.894
Opening	12.190	2.727 (*)	0	0	0	0.919	29.478

There is no change in the preferred model with these two movies removed from the data.

11.87 **a)** The regression model is: $\mu_y = \beta_0 + \beta_1 x_{LOC} + \beta_2 x_{age} + \beta_3 x_{gender}$. **b)** $\beta_0, \beta_1, \beta_2, \beta_3$, and σ. **c)** The estimates are $b_0 = 3.284$, $b_1 = 0.028$, $b_2 = -0.023$, $b_3 = -0.150$, and $s = 0.414$. **d)** $F = 11.008$ with 3 and 96 degrees of freedom. P-value ≈ 0. There is evidence that at least one variable is significant for predicting job stress when LOC, age, and gender are included in the model. **e)** $R^2 = 0.256$. **f)** LOC has a test statistic of 2.271 with df $= 96$, and the P-value $= 0.025$. Age has a test statistic of -4.348 with df $= 96$, and the P-value $= 0.000$. Gender has a test statistic of -1.480 with df $= 96$, and the P-value $= 0.142$. For the model including all three variables, LOC and age are both significant for predicting stress. Gender is not significant.

11.89 **a)** Going from the single-variable model to the full model, the constant changed from 2.2555 to 3.284, the coefficient of LOC decreased from 0.0399 to 0.028, and the estimate of the standard deviation decreased from 0.4513 to 0.414. **b)** 2.598 for the full model and 2.655 for the LOC-only model. **c)** Changing LOC by one unit causes stress level to increase 0.028 in the full model and 0.0399 in the LOC-only model. **d)** $R^2 = 0.256$, which is substantially larger that the LOC-only model, indicating that age and gender are helpful predictors for stress when LOC is also included.

11.91 **a)** Price vs. Promotions: negative, moderate, linear. Price vs. Discount: negative, weak, fairly linear. See the scatterplots below.

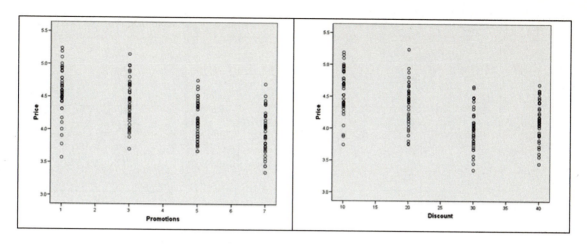

b) The table below shows the mean and standard deviation for expected price for each combination of promotion and discount.

Promotions	Discount			
	10	20	30	40
1	4.920, 0.1520	4.689, 0.2331	4.225, 0.3856	4.423, 0.1848
3	4.756, 0.2429	4.524, 0.2707	4.097, 0.2346	4.284, 0.2040
5	4.393, 0.2685	4.251, 0.2648	3.89, 0.1629	4.058, 0.1760
7	4.269, 0.2699	4.094, 0.2407	3.76, 0.2618	3.780, 0.2144

b) and **c)** At every promotion level, the 10% discount yields the highest expected price, then 20%, then 40%, then 30%. For every discount level, the 1 promotion yields the highest expected price, then 3, then 5, then 7.

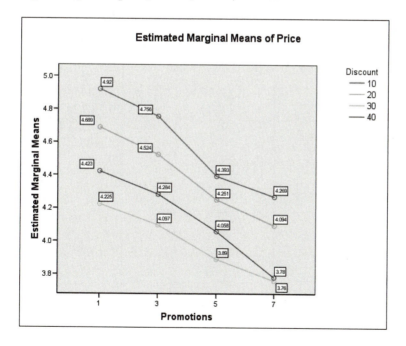

11.93 In the previous exercise, the residual plot for promotions looks okay, but the residual plot for discount seems to have a curved shape. See the plots below.

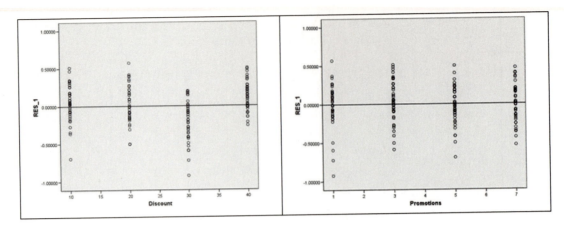

Therefore, try the quadratic term for discount and the interaction of discount and promotion. If the quadratic term for discount is used, the R^2 increases to 61.1%, the s decreases to 0.2511, the F test statistic is 81.809, and the P-value is still very close to 0. All coefficients are significantly different from 0.

If the quadratic term is removed and the interaction of discount and promotion is used instead, the coefficient for the interaction term is not significant, and R^2 and s are very close to what they were in the original model. The best model is the one with the quadratic term for discount: $\hat{y} = 5.540 - 0.102 \times \text{Promotion} - 0.060 \times \text{Discount} + 0.001 \times \text{Discount}^2$.

11.95 a)

	Area	Forest	IBI
Mean	28.29	39.39	65.94
St. dev.	17.714	32.204	18.280
Stemplot	0 . 2	0 . 00000033789	2 . 99
	0 . 5688999	1 . 0014778	3 . 2339
	1 . 0024	2 . 125	4 . 1367
	1 . 66889	3 . 123339	5 . 34556899
	2 . 111133	4 . 133799	6 . 01247
	2 . 66667889	5 . 229	7 . 1112456889
	3 . 112244	6 . 38	8 . 001222344556899
	3 . 9	7 . 599	9 . 1
	4 .	8 . 069	
	4 . 799	9 . 055	
	5 . 244	10 . 00	
	5 . 789		
	2.00 Extremes (>=69)		
	Stem width: 10	Stem width: 10	Stem width: 10
	Right-skewed, 2 high outliers	Right-skewed, no outliers	Left-skewed, no outliers.

at follow. All relationships are linear and moderately weak. Area and forest have a negative relationship, but the others are positive. Data point 40 looks like a potential outlier on the area/forest scatterplot but not anyplace else.

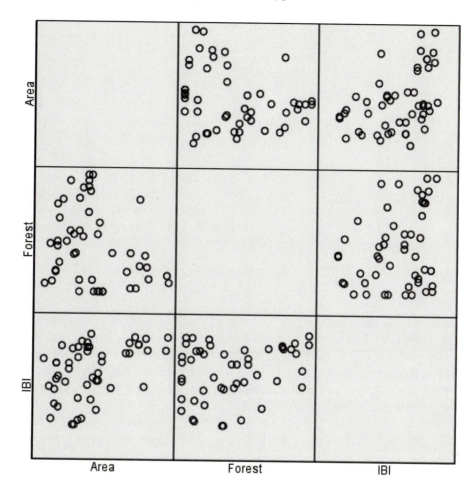

c) $y_i = \beta_0 + \beta_{Area} x_{Area_i} + \beta_{Forest} x_{Forest_i} + \varepsilon_i$. **d)** $H_0 : \beta_{Area} = \beta_{Forest} = 0$, H_a : The coefficients are not both 0. **e)** $R^2 = 35.7\%$, $s = 14.972$, $F = 12.776$, P-value = 0.000. All coefficients are significant. $\hat{y} = 40.629 + 0.569 \times \text{Area} + 0.234 \times \text{Forest}$. **f)** Residual plots look good—random and no outliers. See below.

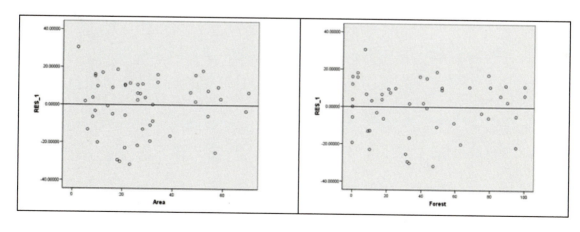

g) A histogram of the residuals looks somewhat skewed left, but the Normal probability plot looks good.

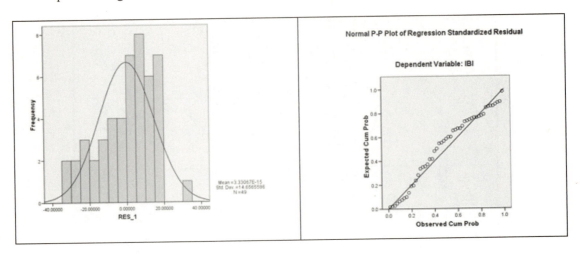

h) Yes. We had general linear relationships between the variables and the residuals look good. However, only 35.7% of the variation in IBI is explained by area and forest.

11.97 **a)** 20%. **b)** The statistically significant variables are strength of expression of WOM with a test statistic of –3.385, amount of WOM given with a test statistic of –3.636, and room for change (PPP) with a test statistic of –16.818. **c)** The more strongly the consumer states their negative opinion about a certain brand, the more negative the impact will be. **d)** When the NWOM is regarding the main brand, the impact will be more negative than when the NWOM is about another brand.

11.99 **a)** The multiple regression equation is $\hat{Y} = 0.906 + 0.027X_1 + 0.211X_2$. The F statistic is 3.152 with a P-value of 0.059. The t test for the coefficient of X_1 has a test statistic of 0.857 with a P-value of 0.399. The test statistic for the coefficient of X_2 is 1.075 with a P-value of 0.292. There is no evidence that either of the explanatory variables has a linear relationship with y when both X_1 and X_2 are in the model. **b)** The model using only X_1 as an explanatory variable yields an F statistic of 5.120 with a P-value of 0.032. The test statistic for the coefficient of X_1 is 2.263 with a P-value of 0.032. X_1 is significant in predicting Y. The regression equation for this model is $\hat{Y} = 0.829 + 0.051X_1$, and the correlation is 0.393. The graph below shows the positive relationship between X_1 and Y.

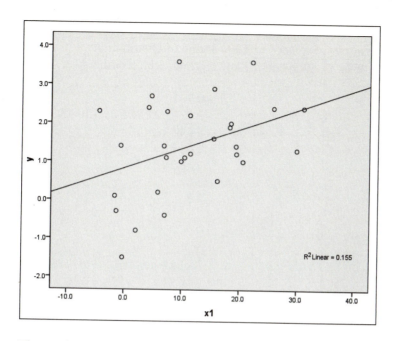

The model using only X_2 as an explanatory variable also shows a positive relationship between the explanatory variable and Y. The regression equation is $\hat{Y} = 1.104 + 0.330X_2$. The correlation is 0.409. An ANOVA F test gives a test statistic of 5.623 with a P-value of 0.025. The t test for the coefficient of X_2 also indicates that this variable is significant in the model with a P-value of 0.025. The scatterplot below shows this relationship.

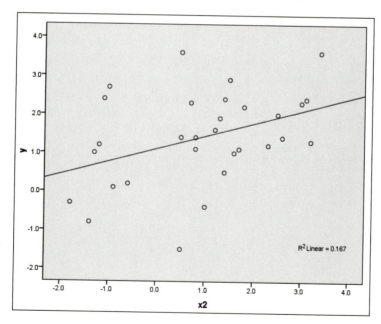

c) Multiple regression results are not sufficient for describing relationships between variables. The relationships between the individual variables should also be considered.

11.101 a)

	Mean	St. Dev.	Description of Distribution

PCB	68.4674	59.3906	Skewed right, 5 high outliers
			
			`0 .   0001111`
			`0 .   2222222223333333333333`
			`0 .   44444455555`
			`0 .   6666677`
			`0 .   888999999`
			`1 .   1111`
			`1 .   23`
			`1 .`
			`1 .   7`
			
			`5.00 Extremes     (>=199)`
			`Stem width:   100.0000`

PCB52	0.9580	1.5983	Skewed right, 6 high outliers
			
			`0 .   0000011111111111`
			`0 .   22222233333333`
			`0 .   4444444555555`
			`0 .   667777`
			`0 .   888899`
			`1 .`
			`1 .   22`
			`1 .   44`
			`1 .   677`
			`1 .   8`
			
			`6.00 Extremes     (>=2.1)`
			
			`Stem width:     1.000`

PCB118	3.2563	3.0191	Skewed right, 5 high outliers
|        |        |        | ```
0 .   234577889
1 .   0113334445555556778899
2 .   1334456679
3 .   133455566889
4 .   0078
5 .   046
6 .   0289

5.00 Extremes     (>=8.2)

Stem width:        1.00
``` |
| PCB138 | 6.8268 | 5.8627 | Skewed right, 5 high outliers |
|        |        |        | ```
0 .   000111
0 .   2222222222223333333333333
0 .   444444555555
0 .   67777
0 .   888888999
1 .   011
1 .   22333

5.00 Extremes     (>=18)

Stem width:       10.00
``` |
| PCB180 | 4.1584 | 4.9864 | Skewed right, 7 high outliers |
|        |        |        | ```
0 .   345667889
1 .   0000111123334578
2 .   111233466678
3 .   011344666779
4 .   1446
5 .   0034
6 .   2
7 .   06
8 .   8
9 .   3

7.00 Extremes     (>=9.5)

Stem width:        1.000
``` |

b) All the variables are positively correlated with each other. All the explanatory variables are significantly correlated with PCB, although PCB52 is the most weakly correlated with PCB and with the other explanatory variables. Only the correlation between PCB52 and PCB180 is not significant (P-value = 0.478).
```

Correlations

		PCB	PCB52	PCB118	PCB138	PCB180
PCB	Pearson Correlation	1	.596**	.843**	.929**	.801**
	Sig. (2-tailed)		.000	.000	.000	.000
	N	69	69	69	69	69
PCB52	Pearson Correlation	.596**	1	.685**	.301*	.087
	Sig. (2-tailed)	.000		.000	.012	.478
	N	69	69	69	69	69
PCB118	Pearson Correlation	.843**	.685**	1	.729**	.437**
	Sig. (2-tailed)	.000	.000		.000	.000
	N	69	69	69	69	69
PCB138	Pearson Correlation	.929**	.301*	.729**	1	.882**
	Sig. (2-tailed)	.000	.012	.000		.000
	N	69	69	69	69	69
PCB180	Pearson Correlation	.801**	.087	.437**	.882**	1
	Sig. (2-tailed)	.000	.478	.000	.000	
	N	69	69	69	69	69

**. Correlation is significant at the 0.01 level (2-tailed).

*. Correlation is significant at the 0.05 level (2-tailed).

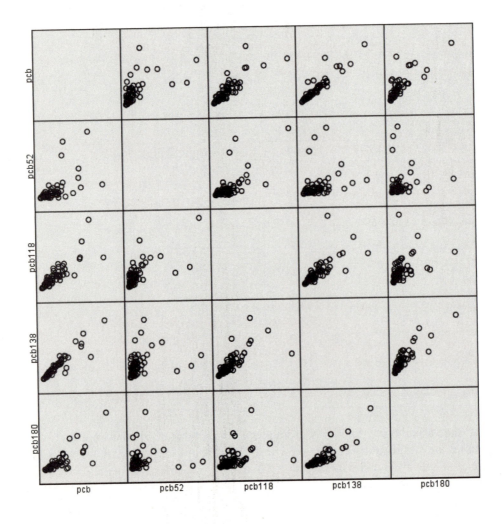

c) See the SPSS output below. $\hat{y} = 0.937 + 11.873 \times PCB52 + 3.761 \times PCB118 + 3.884 \times PCB138 + 4.182 \times PCB180$. $R^2 = 98.9\%$, $s = 6.3821$, $F = 1456.178$, P-value is very close to 0. All the coefficients are significant.

Model Summary[b]

Model	R	R Square	Adjusted R Square	Std. Error of the Estimate
1	.995[a]	.989	.988	6.3820755

a. Predictors: (Constant), PCB180, PCB52, PCB118, PCB138

b. Dependent Variable: PCB

ANOVA[b]

Model		Sum of Squares	df	Mean Square	F	Sig.
1	Regression	237245.8	4	59311.442	1456.178	.000[a]
	Residual	2606.777	64	40.731		
	Total	239852.5	68			

a. Predictors: (Constant), PCB180, PCB52, PCB118, PCB138

b. Dependent Variable: PCB

Coefficients[a]

Model		Unstandardized Coefficients		Standardized Coefficients	t	Sig.	95% Confidence Interval for B	
		B	Std. Error	Beta			Lower Bound	Upper Bound
1	(Constant)	.937	1.229		.762	.449	-1.519	3.393
	PCB52	11.873	.729	.320	16.287	.000	10.416	13.329
	PCB118	3.761	.642	.191	5.855	.000	2.478	5.044
	PCB138	3.884	.498	.383	7.803	.000	2.890	4.879
	PCB180	4.182	.432	.351	9.687	.000	3.320	5.045

a. Dependent Variable: PCB

d) See the histogram of the residuals and the Normal probability plot below. The residuals look approximately Normal.

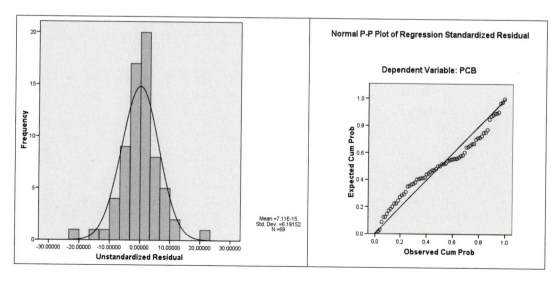

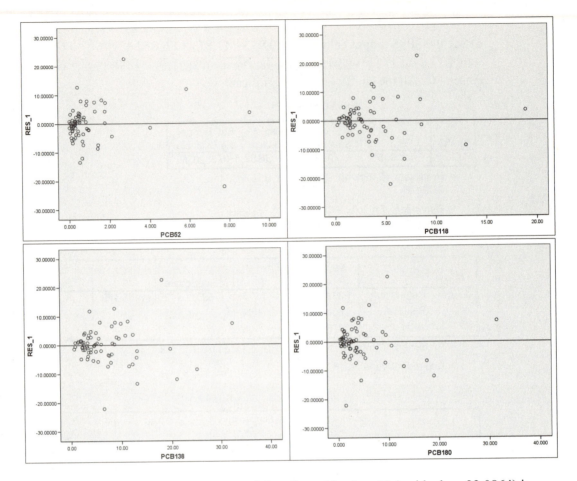

e) Numbers 50 and 65 are the two potential outliers. Number 50 (residual = −22.0864) is the overestimate. **f)** $\hat{y} = 1.628 + 14.442 \times \text{PCB52} + 2.600 \times \text{PCB118} + 4.054 \times \text{PCB138} + 4.109 \times \text{PCB180}$. All coefficients are significant again. $R^2 = 99.4\%$, $s = 4.555$, $F = 2628.685$, and the P-value = 0.

Model Summary[b]

Model	R	R Square	Adjusted R Square	Std. Error of the Estimate
1	.997[a]	.994	.994	4.5553398

a. Predictors: (Constant), PCB180, PCB52, PCB118, PCB138

b. Dependent Variable: PCB

ANOVA[b]

Model		Sum of Squares	df	Mean Square	F	Sig.
1	Regression	218192.6	4	54548.154	2628.685	.000[a]
	Residual	1286.569	62	20.751		
	Total	219479.2	66			

a. Predictors: (Constant), PCB180, PCB52, PCB118, PCB138

b. Dependent Variable: PCB

Coefficients[a]

Model		Unstandardized Coefficients		Standardized Coefficients	t	Sig.	95% Confidence Interval for B	
		B	Std. Error	Beta			Lower Bound	Upper Bound
1	(Constant)	1.628	.886		1.838	.071	-.143	3.398
	PCB52	14.442	.696	.342	20.751	.000	13.051	15.833
	PCB118	2.600	.516	.135	5.034	.000	1.567	3.632
	PCB138	4.054	.375	.407	10.805	.000	3.304	4.804
	PCB180	4.109	.317	.356	12.942	.000	3.474	4.743

a. Dependent Variable: PCB

The residuals look fairly Normally distributed, and the residual plots look random. There are two new potential outliers though at numbers 44 and 58.

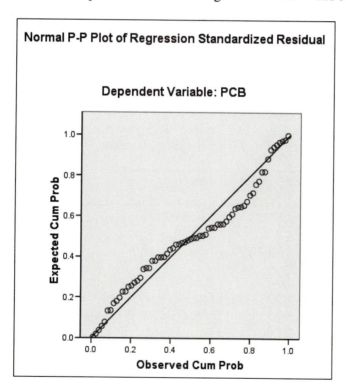

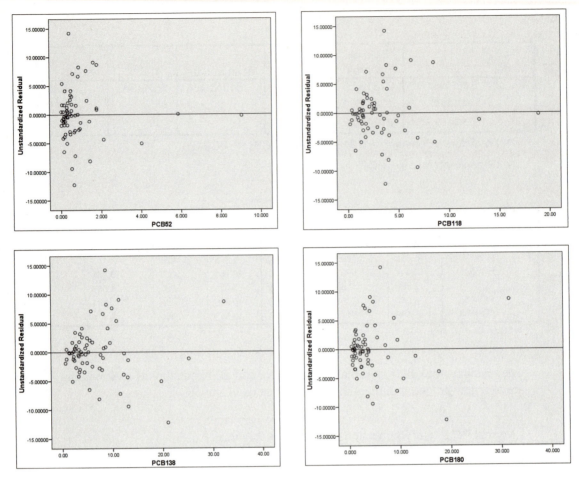

11.103 a) Results will vary with software. SPSS results for logPCB126 are given below.
   b) Most software should ignore these data points, but ignoring data is not a good idea.
   c) Table that follows uses Base 10 logs. Minitab and SPSS give mean = –2.1044 and standard deviation = 0.3325 when using 0.0026 for all the 0 values.

Log of Variable	Mean	St. Dev.	Description of Stemplot
PCB138	0.7009	0.3494	Fairly symmetric, 1 low outlier 1.00 Extremes (=<-.2) –0 . 00 0 . 11233333334444 0 . 5555555555555666666677777788888999999999 1 . 000001112234 1 . 5 Stem width: 1.0000
PCB153	0.7397	0.3914	Fairly symmetric, no outliers –0 . 000 0 . 01222333444444 0 . 5555555555555566667777788888888899999

			1 . 00000011123344 1 . 556 Stem width: 1.0000
PCB180	0.4235	0.4028	Fairly symmetric, 1 high outlier –0 . 000112234 0 . 00000000011112223333333344444444 0 . 55555555566666777778889999 1 . 0122 1.00 Extremes (>=1.5) Stem width: 1.0000
PCB28	–0.5793	0.4918	Fairly symmetric, 1 low and 1 high outlier 1.00 Extremes (=<-2.2) –1 . 6 –1 . 0000011222 –0 . 5555555555666666677777778888899999 –0 . 0001222223333344444 0 . 122333 1.00 Extremes (>=.8) Stem width: 1.0000
PCB52	–0.3354	0.5167	Fairly symmetric, 1 low and 2 high outliers 1.00 Extremes (=<–1.7) –1 . 5 –1 . 112 –0 . 555556666677778888999 –0 . 0000001111112222222333333344444 0 . 1111222234 0 . 67 2.00 Extremes (>=.9) Stem width: 1.0000
PCB126 With 0 values	–1.96	0.228	Right-skewed –22 . 0223568 –21 . 02333444466688 –20 . 146889 –19 . 012359 –18 . 001369 –17 . 00247 –16 . 668

			−15 . 55899 −14 . 9 Stem width: .1000
PCB 126 With 0.0026 for the 0 values	−2.104	0.3325	Fairly symmetric, no outliers. −2 . 5555555555555555 −2 . 2222222 −2 . 00000011111111111111 −1 . 888888999999 −1 . 66677777 −1 . 455555 Stem width: 1.0000
PCB118	0.3717	0.3592	Fairly symmetric, 1 low and 1 high outlier 1.00 Extremes (=<−.6) −0 . 00011234 0 . 000111111111111122222223333344444 0 . 55555555555566666777788999 1 . 1 1.00 Extremes (>=1.3) Stem width: 1.0000
PCB	1.701	0.3483	Fairly symmetric, no outliers 0 . 799 1 . 12233333334444444 1 . 555555555556666666777777788888899999999999 2 . 00000122333 2 . 5 Stem width: 1.0000
TEQ	0.3495	0.2591	Right-skewed, no outliers −0 . 0012 0 . 011245667899 1 . 0333446789 2 . 22345888 3 . 135777 4 . 03488899 5 . 135788 6 . 01358 7 . 001556799 8 . 1 Stem width: .1000

11.105 a) Corn yield = $-41.817 + 4.614 \times$ Soybean yield, $R^2 = 0.909$, $t = 22.151$, and the *P*-value = 0.

b) The residuals appear fairly Normal.

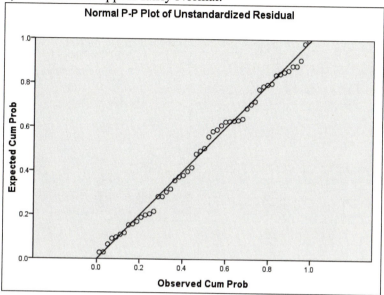

c)

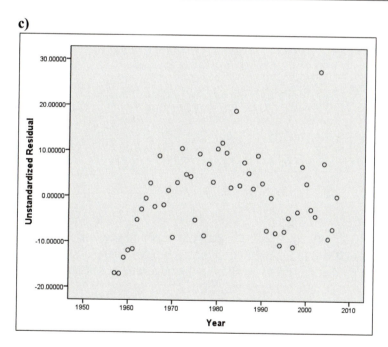

It appears that there is an increasing and then decreasing effect when looking at the residuals plotted against year. It makes sense to include year in the regression model.

11.107 a) Corn yield = $-1508.127 + 0.767 \times$ Year $- 0.014 \times$ Year2 $+ 2.954 \times$ Soybean yield.
b) H_0: All coefficients are equal to zero. H_a: At least one coefficient is not equal to zero. $F = 277.563$, df = 3 and 47, *P*-value = 0. This indicates that the model provides significant prediction of corn yield. **c)** $R^2 = 0.947$ compared to 0.939 from the previous model. **d)** In the order they appear in the model, the *t* statistics for the coefficients are $t =$

4.078 with a *P*-value = 0, $t = -2.501$ with a *P*-value = 0.016, and $t = 6.436$ with a *P*-value = 0. **e)** The residuals appear to be somewhat closer to the zero line when compared to year, but the overall patterns are still apparent.

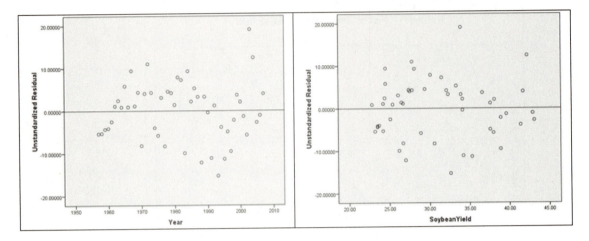

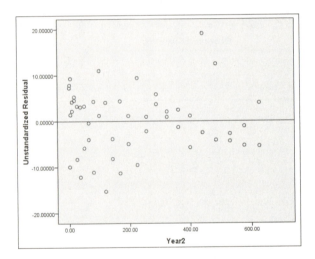

11.109 The prediction interval for 2008 is (131.45, 171.28) when only year is used as an explanatory variable and (129.54, 171.79) when both year and year2 are used. The actual value was contained well within the prediction interval in both cases, but the actual value is closer to the linear prediction of 151.37 than to the quadratic prediction of 150.66.

11.111

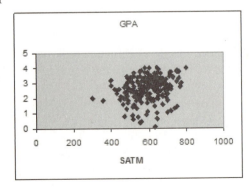

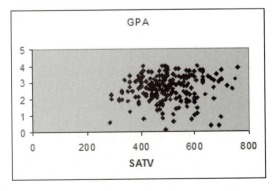

The scatterplots do not show strong relationships. The plot of GPA versus SATM shows two possible outliers on the left side. The plot of GPA versus SATV does not show any obvious outliers.

11.113 The regression equation is GPA = 0.5899 + 0.0343 × HSS + 0.1686 × HSM + 0.0451 × HSE.

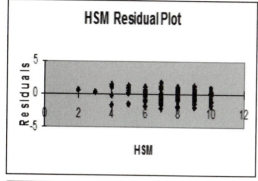

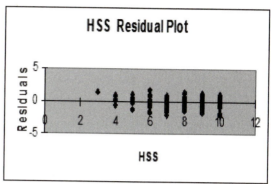

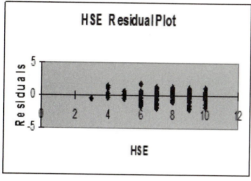

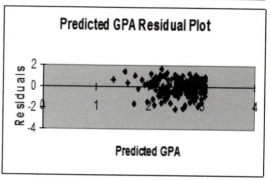

The residuals do not show any obvious patterns.

11.115 a) GPA = 0.666 + 0.193 × HSM + 0.00061 × SATM. **b)** H_0: All coefficients in the regression model are zero. H_a: At least one coefficient is not zero. This means that we assume the model is not a significant predictor of GPA and let the data provide evidence that the model is a significant predictor. $F = 26.63$ and the P-value = 0. This model provides significant prediction of GPA. **c)** HSM: (0.1295, 0.2565) SATM: (−0.00059, 0.001815). Yes, the interval describing the SATM coefficient does contain zero. **d)** HSM: $t = 5.99$, P-value = 0. SATM: $t = 0.999$, P-value = 0.319. Based on the large P-value associated with the SATM coefficient one should conclude that SATM does not provide significant prediction of GPA. **e)** $s = 0.703$. **f)** $R^2 = 0.1942$.

11.117 Looking at the sample of males only shows the same results when compared to the sample of all students. $R^2 = 0.184$ (compared to 0.20) and, while HSM is a significant predictor, HSS and HSE are not.

11.119 a) GPA = 0.582 + 0.155 × HSM + 0.050 × HSS + 0.044 × HSE + 0.067 × Gender + 0.05 × GHSM − 0.05 × GHSS − 0.012 × GHSE. The t statistics for each coefficient have P-values greater than 0.10 for all except the explanatory variable HSM. **b)** Verify. **c)** Verify. **d)** $F = 0.143$, P-value = 0.966. This indicates there is no reason to include gender and the interactions.

Case Study 11.1

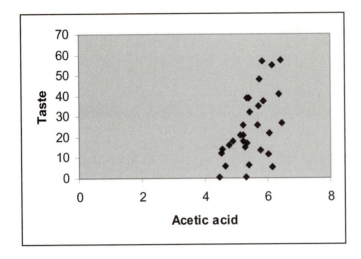

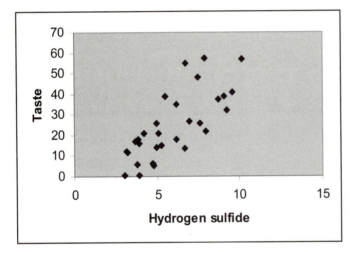

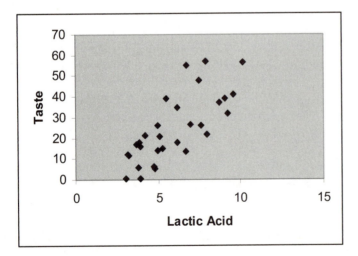

The relationships between taste and hydrogen sulfide and taste and lactic acid appear slightly stronger than the relationship between taste and acetic acid. The correlation coefficient values support this. The values for *r* are 0.7558, 0.7042, and 0.5500, respectively. After running the analysis with interaction effects, it was found that the strongest model in terms of R^2 was the model with both the hydrogen sulfide term and the lactic acid term. Interaction did not seem to significantly improve the model's predictive ability. The best model for predicting taste would be: $\hat{y} = -27.50 + 3.95 \times$ hydrogen sulfide $+ 19.89 \times$ lactic acid.

Case Study 11.2

The model chosen as the best predictor of GPA includes the variables IQ, SC, C1, C3, and C5. The original analysis included all 10 variables. Reviewing the residuals scores and residual plots showed that there were three outliers: cases 55, 52, and 22. After removing these cases and rerunning the analysis, keeping with only the variables that appeared significant at a 5% level, the resulting model is: $\hat{y} = -3.706 + 0.0775 \times$ IQ $+ 0.1632 \times$ C1 $+ 0.0942 \times$ C3. $R^2 = 0.61$ and standard error $= 1.156$. There does not appear to be any reason to question the regression assumptions.

Case Study 11.3

The mean earnings for females is \$20,708.56, and the mean earnings for males is \$20,569.16. The side-by-side boxplots look fairly similar for both genders with many high outliers. The average earnings for part-time workers is \$19,033.25, and the mean earnings for full-time workers is \$21,291.12. The side-by-side boxplots for status show that the distribution of full-time workers is higher than that of part-time workers.

The correlation between earnings and gender is -0.012 and is not significant. The correlation between earnings and status is 0.224 and is significant. Gender is also significantly correlated with status ($r = 0.081$).

The ANOVA F test concludes that at least one of the explanatory variables will be able to tell us something about salary. With both gender and salary included in the model, the equation is $\hat{y} = 19077.147 - 354.928 x_{gender} + 2282.401 x_{status}$ where $x_{gender} = 1$ for male and 0 for female, and $x_{status} = 1$ for FT and 0 for PT. R^2 is 5.1% (which is very low), and the standard error of the estimate is 4361.671. The coefficient for gender is not significantly different from 0.

Using just status, the model is $\hat{y} = 19033.254 + 2257.867 x_{status}$. R^2 didn't change much and is now 5.0%. The standard error of the estimate has stayed much the same at 4362.486. The coefficient of status is significantly different from 0.

The model that uses just status is the better model. Therefore, gender is not significantly related to salary for hourly employees.

Case Study 11.4

Both job level ($r = 0.798$) and gender ($r = 0.157$) are significantly correlated with salary. Job level and gender are also significantly correlated with each other ($r = 0.141$).

Using both job level and gender to predict salary, the model where $x_{gender} = 1$ for male and 0 for female is $\hat{y} = -18113.2 + 6114.266 x_{joblevel} + 1278.652 x_{gender}$ with an R^2 of 63.9%

and a standard error of the estimate of 8391.326. The coefficient for gender is not significant.

With only job level, with gender dropped from the model, the equation of the line is: $\hat{y} = -17905.7 + 6163.957 x_{joblevel}$. The new R^2 is 63.7%, and the new standard error of the estimate is 8404.060. There is not much of a change in either of these values, and the coefficient for job level is still significant. Therefore, the model using only job level is the better model. Gender is not a good predictor variable for salary.

Case Study 11.5

The zip code chosen for analysis was 47905. There were 158 homes in this zip code. After looking at the data, four homes were removed as having very high square footage. These were observations 453, 468, 485, 500, and 504. One additional observation, observation 502, was removed as an outlier for price. The scatterplot that follows shows the linear trend between price and square footage with the outliers removed. The residual plots did not show any reason to believe the relationship was not linear.

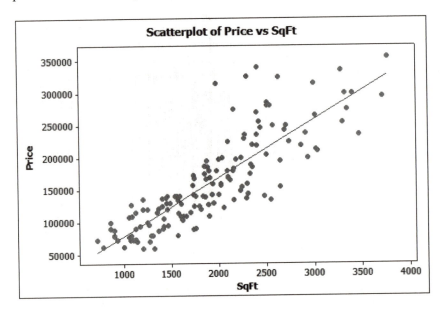

Further analysis was done with the bedroom and bathroom variables. It was found that the number of bedrooms was not a significant predictor of price. This is consistent with the results found in case 11.3. The number of bathrooms did contribute significantly to the price of a home. The indicator variables used were Bath25 and Bath3. Bath25 was equal to 1 if the home had 2.5 bathrooms and 0 otherwise. Bath3 was equal to 1 if the home had more than 2.5 bathrooms and 0 otherwise. When Bath25 and Bath3 were both zero this indicated that the home had either one or two bathrooms. The model with SqFt, Bath25, and Bath3 gave the best fit.

$\hat{y} = \$7027 + 72.239 \times \text{Sqft} + 29{,}572 \times \text{Bath25} + 38{,}439 \times \text{Bath3}.$
$R^2 = 0.741$ and standard error $= 36{,}131.3.$

Chapter 12: Statistics for Quality: Control and Capability

12.1 Answers will vary. Check for reasonableness and completeness of process as well as understanding of and format of cause-and-effect diagram.

12.3 Answers will vary. Check for reasonableness, understanding of decision points, format of flowchart, and completeness. Some possible causes might be size of picture, time of day, speed of Internet connection, etc.

12.5 Common cause variation would include longer lines at security checkpoints or position in line for boarding the plane. Special cause variation would include forgetting an item such as a passport and having to return to the vehicle to retrieve it, a delayed plane, or a power outage at the airport.

12.7 Answers will vary.

12.9 Answers will vary.

12.11 A point that falls above the R chart's upper limit signals that a subgroup has greater variability than is expected based on the set of subgroups being examined. However, a point falling above the UCL for the $\bar{x}$ chart indicates that the average for a single group is greater than expected and that the process is out of control.

12.13 After removing the two observations as discussed in the text, verify using $D_4 = 2.282$, $D_3 = 0.000$ and $\bar{R} = 0.00899$ for the R chart. Verify the $\bar{x}$ chart using $A_2 = 0.729$ and $\bar{\bar{x}} = 2.60755$.

12.15 Verify the s chart using $B_4 = 2.089$, $B_3 = 0.000$, and $\bar{s} = 9.2015$. For the $\bar{x}$ chart, verify using $A_3 = 1.427$ and $\bar{\bar{x}} = 40.62$.

12.17 Answers will vary.

12.19 Specifications limits start with the customer. When the customer states their product or service requirements, the product/service designer translates those requirements into product/service specifications. These specifications define limits within which the product can function as intended. Control limits, on the other hand, are derived from process behavior. They are independent of product/service specifications. Control limits tell us what the expected variability is in the process with respect to certain product, process, or service characteristics.

12.21 $P(\text{diameter} \leq 2.592) + P(\text{diameter} \geq 2.632) = 0.000086$. This would be 86 parts per million that are defective.

12.23 $P(\bar{x} > 713 \mid \mu = 693) + P(\bar{x} < 687 \mid \mu = 693) = 0.0004 + 0.1587 = 0.1591$.

12.25 $\bar{R} = 0.372$, so the R chart control limits are LCL $= 0$ and UCL $= 0.7864$. $\bar{\bar{x}} = 10.5812$, resulting in an LCL of 10.0.3666 and a UCL of 10.7958 for the $\bar{x}$ chart.

12.27 Assuming independent observations, there would be a $(0.5)^9 = 0.001953$ chance that nine successive observations would be on the same side of the CL.

12.29 In the chart that follows, we see that the last nine observations are all below the CL, indicating that the process is out of control based on the nine-in-a-row rule. Subgroups 27 through 35 are affected with the out-of-control signal coming at subgroup 35.

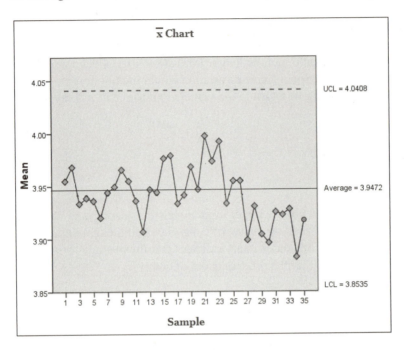

12.31 The s chart has a lower control limit of 2.1318 and an upper control limit of 12.8809. The lower control limit of the $\bar{x}$ chart is 23.3646 and the upper control limit is 38.002.

12.33 Answers will vary.

12.35 **a)** 97.43%. **b)** $\hat{C}_{pk} = 0.73$.

12.37 Answers will vary.

12.39 A process is said to be at six-sigma quality when the distance between the mean of the process and each of the specification limits is equal to six times the standard deviation of the process.

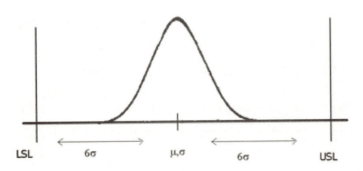

12.41 **a)** 28750, $\bar{p} = 0.0334$. **b)** CL = 0.0334, UCL = 0.0435, LCL = 0.0233.

12.43 CL = 6.3043, LCL = −1.23 → 0, and UCL = 13.84. No remaining counts are outside these new control limits.

12.45 CL = 0.006, UCL = 0.0131, LCL = 0.

12.47 **a)** $\bar{p} = 0.3555$, $\bar{n} = 922$. **b)** UCL = 0.403, LCL = 0.308. The process appears to be in control.

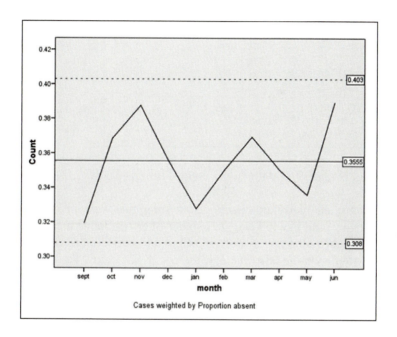

c) Adding exact limits does not affect the conclusions. The process remains in control. For October, UCL = 0.402 and LCL = 0.309. For June, UCL = 0.404 and LCL = 0.307.

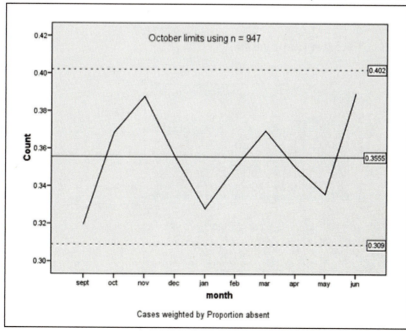

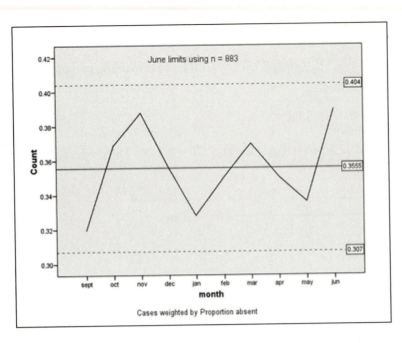

12.49   The upper control limit is 26.619, and the lower control limit is 3.381.

12.51   Often with attribute control charts we are counting something that we want to keep under control. For example, we would like to keep the number of errors below a certain number. Therefore, a point falling above the upper control limit would indicate that the item we are counting occurred too often. On the other hand, if the count was below the lower control limit, then we would not be concerned as this would mean that we have fewer of the problems we were counting than what we expected. We should investigate the cause and use it to improve the process if possible.

12.53   **a)** The percents do not add up to 100 because some customers have more than one complaint.

b)

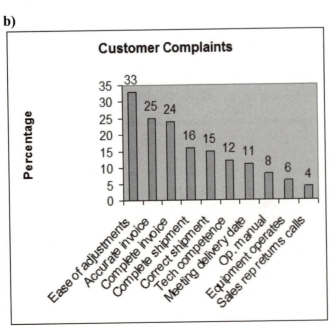

Choose "Ease of adjustments" for focusing your improvement efforts.

12.55   CL = 7.65, UCL = 19.65, LCL = 0. Sample number 1 is above the UCL and sample number 10 is very close to the UCL.

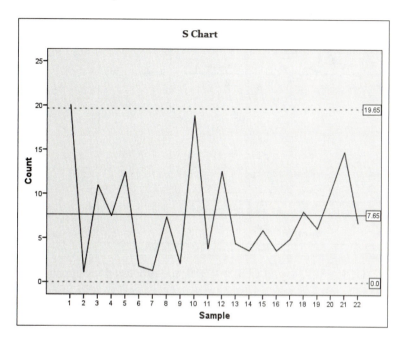

12.57   **a)** A p chart would be used with UCL = 0.0194 and LCL = 0. **b)** It is very unlikely that we will observe unsatisfactory films in a sample of 100.

12.59   **a)** The s chart is in control with center line = 0.0028, UCL = 0.0092 and LCL = 0.

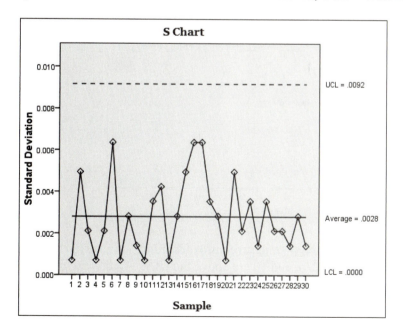

b) The $\bar{x}$ chart is also in control with center line = 1.2619, UCL = 1.2693 and LCL = 1.2544.

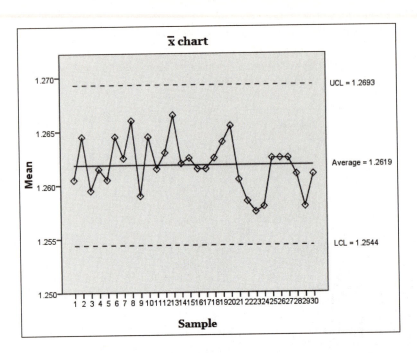

12.61   a) With $\bar{p} = 0.522$, UCL = 0.734, and LCL = 0.310, the process does appear to be in control.

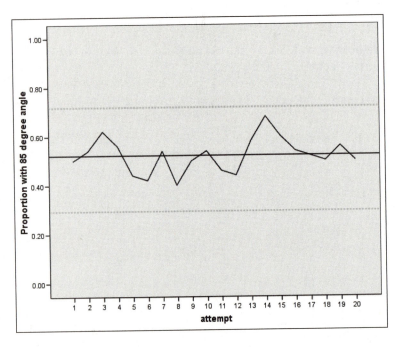

b) The new sample proportions are not in control according to the original control limits from part (a). Attempts 2, 4, and 10 are above the UCL. With new $\bar{p} = 0.702$, the new limits for future samples should be UCL = 0.896 and LCL = 0.508.

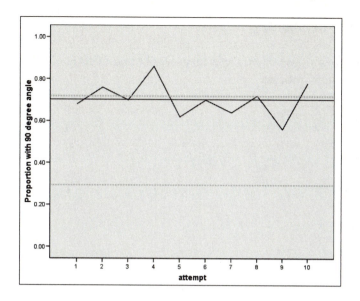

12.63 **a)** Using a c_4 of 0.99363, the LCL is −10.98, and the UCL is 12.49.

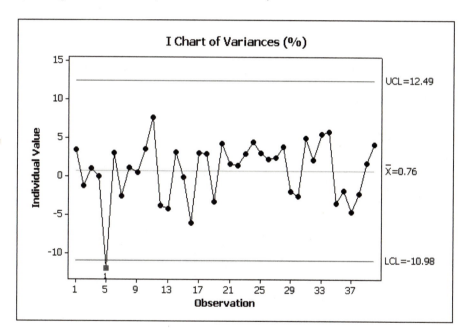

The 5th observation is outside the lower control limit. This is the point with a variance of −11.98. **b)** The 20th through the 28th observations are all above the center line. Based on the nine-in-a-row rule, this would result in an out-of-control signal. **c)** The new control limits are LCL = −10.32 and UCL = 11.38. There are no further out of control signals, so these limits could be used for future monitoring.

Chapter 13: Time Series Forecasting

13.1   **a)** The time series plot below shows that there is an upward trend over time, so this series is not consistent with a random process.

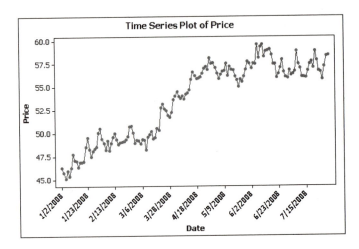

   b) There are two runs in this series which is consistent with the previous conclusion.

13.3   There are 42 observations above the mean and 48 below the mean. Thirty-five runs are observed, but the expected number of runs is 45.80 with a standard deviation of 4.6955. This results in $Z = -2.30$ and a P-value of 0.0214.

13.5   **a)** 78. **b)** P-value $= 0.1630$.

13.7   **a)** $t = 7.97$. P-value $= 2.746E\text{-}09$. There is evidence that the regression coefficient for the trend line is not zero. **b)** Predicted sales are 314.91 thousand feet for January 2008 and 319.44 thousand feet for February 2008.

13.9

Month (2008)	September	October	November	December
Forecasted Sales (millions)	$46,393.57	$49,167.45	$55,745.32	$72,329.57

13.11 $\log(\widehat{Cars_t}) = 3.00 + 0.266969t$.

13.13   Since the amount of increase or decrease appears to be relatively constant from year to year for a given month, it would be more appropriate to use additive seasonality.

13.15   **a)** 1.3658. **b)** When Q1, Q2, and Q3 are set to zero, we are left with the constant 715.91397 times the trend component, which is the prediction equation for the fourth quarter. Since the trend components for the trend-only model and the trend-and-seasonal model are nearly equal, the factor of 1.3658 implies fourth-quarter sales are 1.3658 times the general trend of the series.

13.17 a)

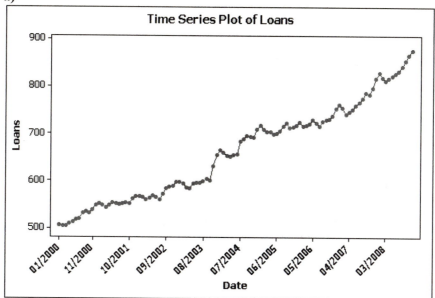

b) The data appear to have a somewhat curved upward trend. **c)** The data do exhibit seasonality. There is a peak in January of each year, followed by a decrease for the next couple of months, then a fluctuating increase up to the next January. **d)** There are three groups of points seen in these data. The first is through approximately January 2004. The second is from January 2004 to January 2007, and the third begins at January 2007. In the first of these three groups, the level of consumer loans is increasing gently. Then there is a sizeable increase at the end of 2003. At this time, we see a higher level of loans with more fluctuation in the loan amounts. Finally, starting at the beginning of 2007, we can see much steeper increases occurring throughout the two years shown in the data.

13.19 a)

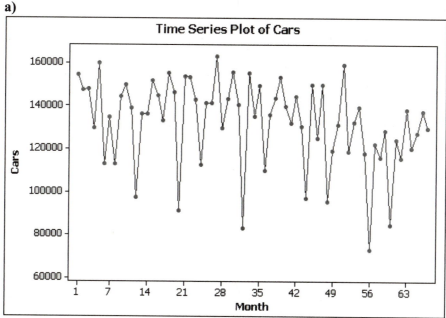

b) There may be a slight decreasing trend in the series as seen in the plot above.
c) August and December exhibit lows in every year.

13.21 **a)** The series has a downward trend that gets steeper as the years increase. From 1981 to 1991, there is only a slight downward trend, but the values start decreasing quite rapidly after 1991.

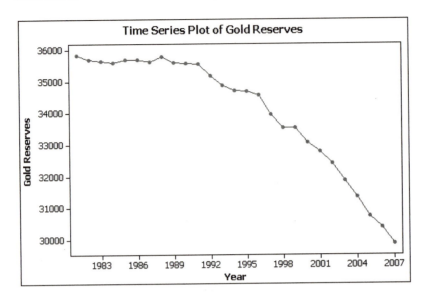

b) There are two observed runs in the data while the expected number of runs is 14.04. The standard deviation is 2.457, resulting in a test statistic of $z = -4.90$. The P-value is approximately zero. There is evidence that this process is not random. **c)** Over time, the distances between successive observations are getting larger.

13.23 **a)**

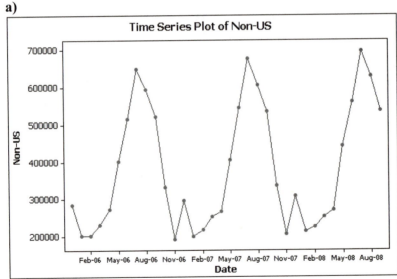

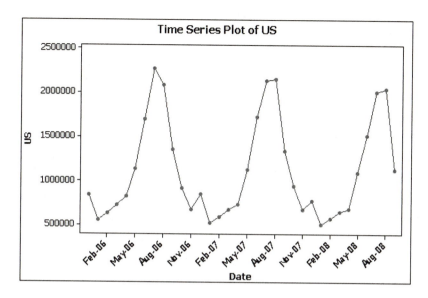

b) For the non-U.S. visitors, July is the month with the highest number of visitors, followed by August, then June and September. For U.S. visitors, July and August have the highest numbers (July for 2006, August for 2007 and 2008), and June comes in third. For the off-peak season, December shows a bit of a surge in both U.S. and non-U.S. visitors. **c)** There appears to be slight positive trend for the non-U.S. data and a slight negative trend for the U.S. data. Software will likely show a positive trend for both data sets due to ending the series with the values for the higher months; however, looking at the results for particular months, we can see the trends more clearly.

13.25 **a)** This is a decreasing series as seen below.

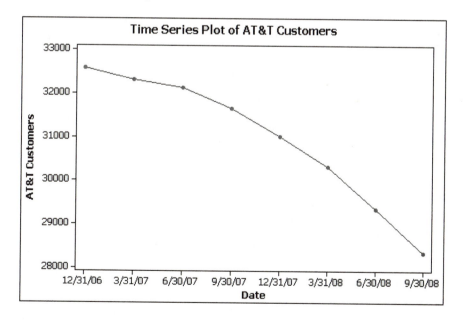

b) There is no apparent seasonality in this series. **c)** $\hat{y}_t = 33682.6 - 604.643t$. **d)** As seen in the plot below, the linear model is neither adequate for describing this series nor for predicting future values. Notice that the prediction for any future values would be too high based on this model.

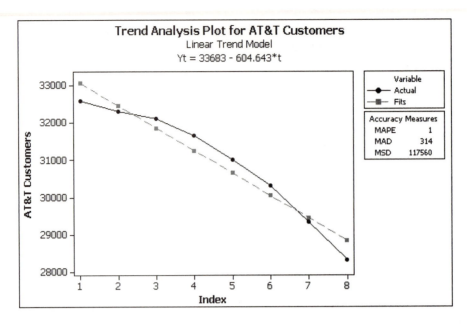

13.27 **a)** (−1.20, 1.37). **b)** (57.18, 59.75).

13.29 **a)** Based on the time series plot below, there is no evidence of a trend over time.

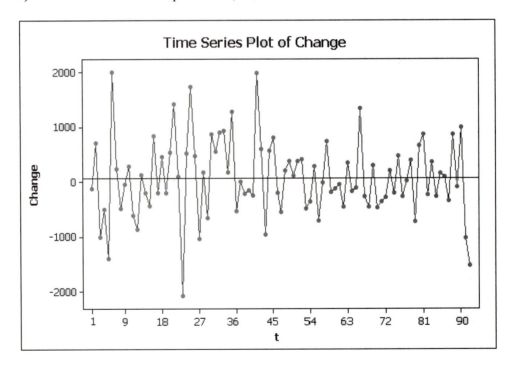

b) While no months always fall on one side of the sample mean line or the other, June and July are the most consistent with June falling below the line 7 out of 8 times and July falling above the line 7 out of 8 times. January and December also tend to fall below the line and October above the line (5 out of 7 times). **c)** $\hat{y}_t = -188.6 + 493.9Feb +$ $230.7Mar - 36.3Apr + 329.5May + 72.8Jun + 975.1Jul - 73.2Aug + 129.5Sep + 798.9Oct$ $+ 293.9Nov - 115.3Dec.$ **d)** $\hat{y}_{93} = -59.1, \ \hat{y}_{94} = 610.3.$

13.31 a)

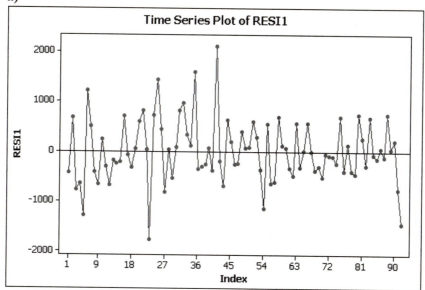

The residuals appear to have a random pattern with some short runs. **b)** The observed number of runs is 45, and the expected number of runs is 46.9130. The *P*-value of the runs test is 0.688. There is not enough evidence against randomness to say that the residuals are nonrandom. **c)** The correlation is approximately 0, and the lagged residual plot shows random scatter in all four quadrants, so there is no evidence of autocorrelation.

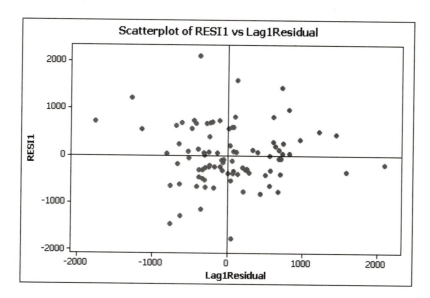

13.33 The lagged residual plot will show points with a strong positive association for S1, a strong negative association for S2, and no particular association with points spread among all four quadrants for S3.

13.35 $\hat{y}_t = 230 + 10t - 30Q_4$.

13.37 a)

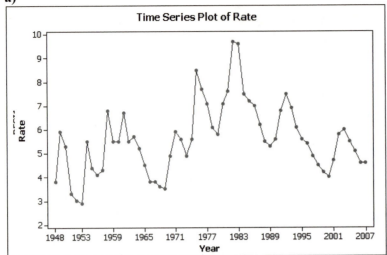

b) There appears to be a slight upward trend with many local high points. The line appears to follow a snake-like path with high values occurring in groups and low values occurring in groups. **c)** There is a positive linear relationship between Rate and Rate lagged one period, suggesting that an AR(1) model is appropriate.

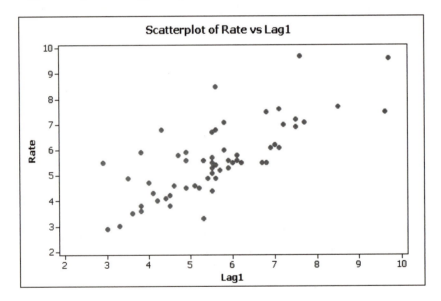

13.39 $\hat{y}_{37} = 5337.01.$

13.41 a)

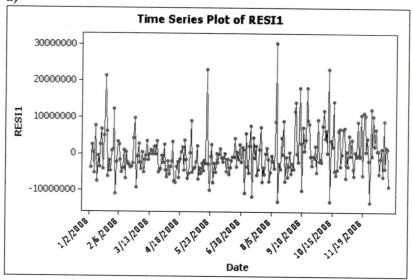

The residual series exhibits less meandering behavior than the original series.

b)

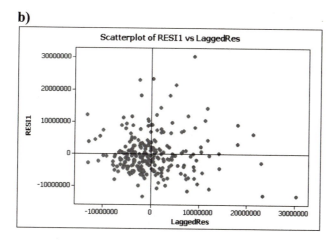

This plot does not have the strong linear trend seen in the previous exercise. The residuals appear to have a pretty random scatter, so the indication is that the autocorrelation has been accounted for. **c)** The histogram and probability plot below both show that the distribution of the residuals is not Normal but skewed to the right.

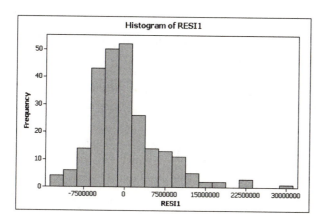

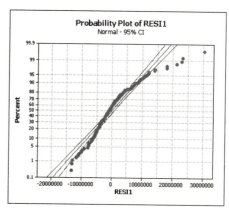

13.43   **a)**

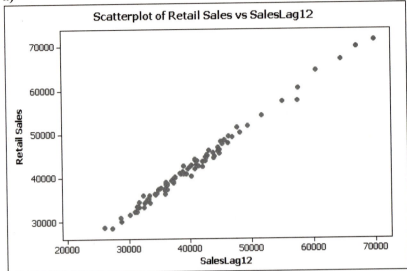

b) $R = 0.995$.   **c)** There is a very strong, linear relationship between the 12-month lagged variable and retail sales, so the 12-month lagged variable does appear to be a good predictor.

13.45   **a)**

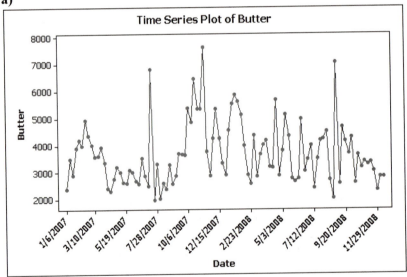

There does not appear to be a trend in these data or any significant shifts.   **b)** 1-week, 2-week, 3-week, 4-week, 5-week, and 6-week moving averages are 2835.91, 2835.04, 2667.33, 2758.07, 2877.52, and 2945.09.   **c)** Below are the moving average residuals for one point. The remaining residuals should be calculated equivalently.

	1-wk	2-wk	3-wk	4-wk	5-wk	6-wk
Residual for 12/13/08	1.73	252.87	103.78	−52.02	−131.02	−200.62

13.47 a)

	1-wk	2-wk	3-wk	4-wk	5-wk	6-wk
MSE for moving avg	1,929,978	1,465,070	1,319,551	1,316,952	1,292,744	1,342,976

b)

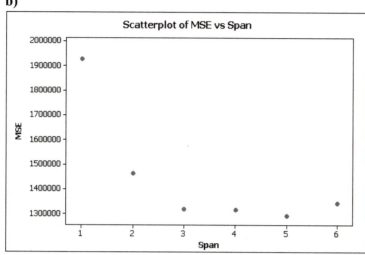

The MSE decreases sharply at first and then gradually to a low value at a span of 5. At a span of 6 the MSE increases.

13.49   a) The forecast for 2008 is 315.80, based on the following predicted values:

Year	1996	1997	1998	1999	2000	2001
$\hat{y}_{year}$	292.31	289.24	288.83	292.93	299.17	307.16

Year	2002	2003	2004	2005	2006	2007
$\hat{y}_{year}$	309.69	310.27	311.31	310.23	309.63	313.37

b) The forecast for 2008 is 325.26, based on the following predicted values:

Year	1996	1997	1998	1999	2000	2001
$\hat{y}_{year}$	292.31	280.01	285.76	304.62	320.23	335.33

Year	2002	2003	2004	2005	2006	2007
$\hat{y}_{year}$	322.94	314.64	315.33	307.79	307.33	324.13

c) $\hat{y}_{2009} = 0.2(y_{2008}) + 0.8(315.80)$.

13.51

a)	b)	c)
0.1000	0.5000	0.9000
0.0900	0.2500	0.0900
0.0810	0.1250	0.0090
0.0729	0.0625	0.0009
0.0656	0.0313	9E-05
0.0590	0.0156	9E-06
0.0531	0.0078	9E-07
0.0478	0.0039	9E-08
0.0430	0.0020	9E-09
0.0387	0.0010	9E-10

d)

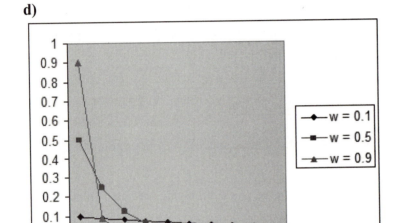

e) The curve that starts the highest, at $w = 0.9$, puts more weight on the most recent value of the time series. **f)** 0.0349, 0.0005, 9×10^{-11}. These are smaller than the first ten coefficients. $w = 0.1$ puts the most weight on y_1.

13.53 **a)**

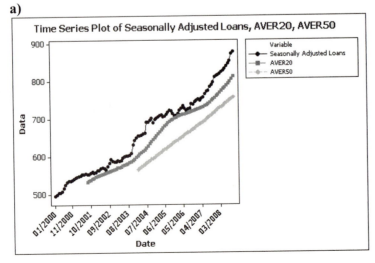

b) As the span increases, the points in the moving average decrease, getting farther away from the actual values. Since the series is increasing, moving averages with longer spans will include more low values and will thus not do a good job of approximating the actual time series. **c)** The results are consistent with the quote. Neither of the moving averages shown here would be appropriate for predicting future values as both would substantially underestimate these values. However, in both of the moving averages, we can see the trend of the overall data.

13.55 **a)** $\hat{y}_t = wy_{t-1} + (1-w)\hat{y}_{t-1} = wy_{t-1} + \hat{y}_{t-1} - w\hat{y}_{t-1} = \hat{y}_{t-1} + w(y_{t-1} - \hat{y}_{t-1}) = \hat{y}_{t-1} + we_{t-1}$.

b) The forecasted value is equal to the predicted value for the previous observation plus a portion of how far this predicted value was from being accurate.

13.57 **a)** $\hat{y}_t = 2964 + 298.26t$. $R^2 = 0.685$, and $s = 5429.71$. **b)** $\hat{y}_t = 11{,}438 - 242.65t + 5.8163t^2$. $R^2 = 0.832$, and $s = 3987.92$. **c)** $\hat{y}_t = 6775 + 343.2t - 9.847t^2 + 0.11228t^3$. $R^2 = 0.862$ and $s = 3637.94$. **d)** The third-degree polynomial is the best model of these three. In this final model, all three variables are significant (P-values of 0.021, 0.008, and 0.000), the R^2 value is the highest of the models, and the regression standard error is significantly lower than with either of the other models.

13.59 **a)**

b)

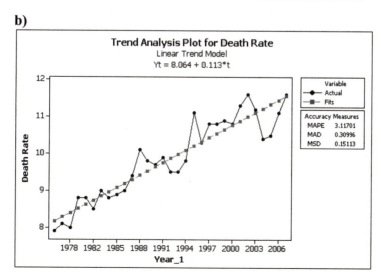

The model is $\hat{y}_t = 8.064 + 0.113t$. **c)** The death rate is increasing by 0.113 per year on average. **d)** For 2008, the estimated rate of death loss is 11.68.

13.61 **a)** The precipitation data appears to be approximately Normally distributed.

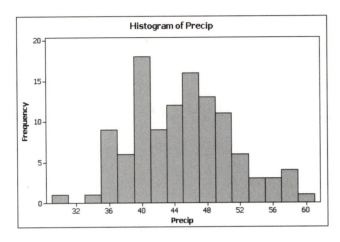

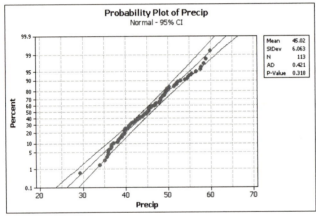

b) (35.44, 55.51).

13.63 a)

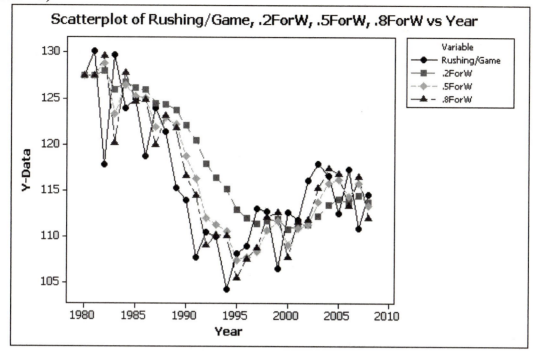

b) The 2009 predictions are 113.91 for $w = 0.2$, 113.98 for $w = 0.5$, and 114.08 for $w = 0.8$. The prediction models are smoother as w gets smaller.

13.65   Using the 3-year moving average, $\hat{y}_{94} = 39,818$. Using the 15-year moving average, $\hat{y}_{94} = 34,270$.

13.67   **a)** MAD = 16.919, MSE = 550.54, and MAPE = 5.773%. **b)** MAD = 12.661, MSE = 272.26, and MAPE = 4.589%. **c)** The AR(1) model is better based on these measures since lower numbers here are more desirable.

13.69   **a)** $\hat{y}_t = 298,728 - 511.8t - 6616Jan - 41,143Feb - 41,041Mar + 17,280Apr - 42,361May - 46,355Jun - 28,947Jul - 70,970Aug - 21,806Sep + 36,377Oct - 5378Nov.$

b)

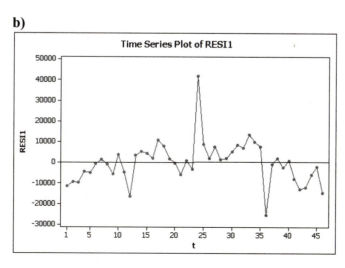

The time series plot shows that the residuals have an increasing trend followed by a decreasing trend instead of random scatter, which indicates that there is evidence of nonlinearity. **c)** $\hat{y}_t = 281{,}883 + 1224.4t - 36.941t^2 - 2774Jan - 37{,}597Feb - 37{,}716Mar + 20{,}457Apr - 39{,}258May - 43{,}252Jun - 25{,}771Jul - 67{,}646Aug - 18{,}260Sep + 40{,}219Oct - 5378Nov$. January has an estimated coefficient of -2774 with a *P*-value of 0.721.

13.71 **a)** $\hat{y}_t = 0.00476 + 0.90026y_{t-1} + 0.012436 \times \text{Ind}_{2005}$. **b)** The coefficient has increased from 0.86099 to 0.90026. **c)** The 2008 prediction is $\hat{y}_t = 0.05086$. This estimate is a little lower than the Example 13.14 prediction which was 0.0511286.

13.73 **a)**

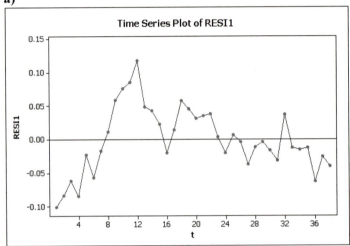

The time series plot of the residuals shows an increasing then decreasing pattern.

b)

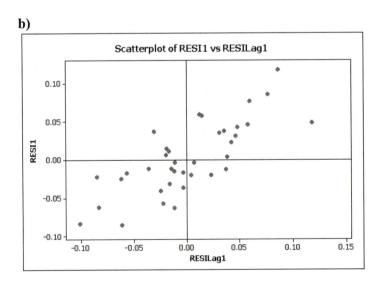

The lagged residual plot does show evidence of autocorrelation. The correlation between e_t and e_{t-1} is 0.766. **c)** $\widehat{\text{Log}(y_t)} = 2.4843 + 0.008629t - 0.69311Q1 - 0.29161Q2 - 0.25568Q3 + 0.7338\,\text{Log}(y_{t-1})$.

d)

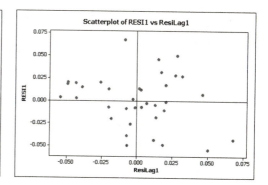

The residuals here do not exhibit a pattern, and the lagged residual plot shows random scatter in all four quadrants. These support the adequacy of the fitted model. **e)** The forecast is 11,036, which is lower than the previous forecast of 11,545. This new forecast seems more in line with what we would expect to see in our original time series.

13.75 a)

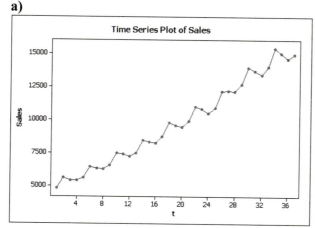

The time series has an increasing trend with obvious seasonality. The second quarter has a sizable jump each year from the first quarter. Sales then decrease in the third quarter and again in the fourth quarter, rising slightly in the first quarter. **b)** $\hat{y}_t = 4158 + 296t$.

c) $\hat{y}_t = 4706.0 + 211.64t + 2.2174t^2$. **d)** $\hat{y}_t = 5071.88e^{0.031605t}$.

e)

	MAD	MSE	MAPE
Linear Model	362	203414	4.014%
Quadratic Model	318	152408	3.315%
Exponential Trend Model	373	233990	3.826%

All three measures are lowest for the quadratic model, so this one provides the best fit.

13.77   a)

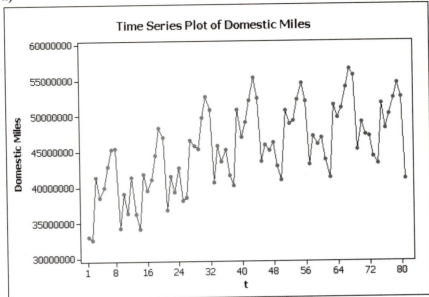

The time series appears to be curved. In general, the trend of the number of domestic miles increases through about the middle of 2005 when it levels out. It begins to decrease toward the end of 2007. **b)** The seasonal variation appears additive as the increases and decreases for particular seasons are not changing significantly as the series progresses. **c)** $\hat{y}_t = 35,400,927 + 364,870t - 2671.2t^2 - 3,706,652Jan - 5,175,343Feb + 3,752,082Mar + 1,168,197Apr + 2,214,362May + 5,242,785Jun + 7,849,675Jul + 6,118,787Aug - 4,219,674Sep + 139,004Oct - 1,816,936Nov$. The estimated domestic miles for October 2008 are 47,499,467.

13.79   a)

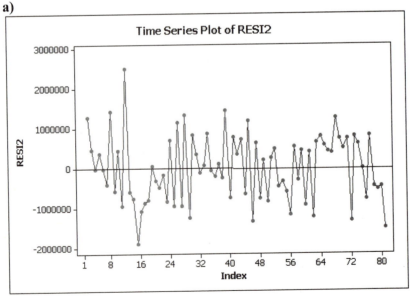

The time series of the residuals now seems to bounce back and forth across the zero line with most runs having only a length of 1.

b)

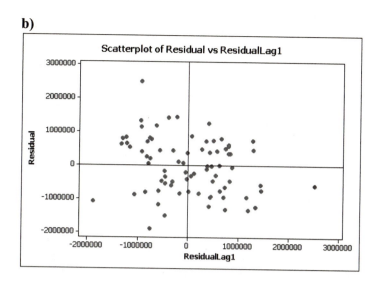

The residual and the lag one residual have a correlation of –0.24. The lagged residual plot also shows a negative relationship between these two. There does still appear to be some evidence of autocorrelation in the series that we have not accounted for in the current model. **c)** $\hat{y}_t = 10{,}146{,}180 + 120{,}623t - 1049.1t^2 - 3{,}558{,}944Jan - 4{,}783{,}694Feb + 6{,}455{,}376Mar + 1{,}860{,}862Apr - 174{,}651May + 3{,}656{,}518Jun + 4{,}916{,}885Jul + 1{,}111{,}092Aug - 9{,}836{,}686Sep - 1{,}908{,}367Oct - 675{,}931Nov + 0.2959y_{t-1} + 0.4307y_{t-2}$. Both y_{t-1} and y_{t-2} are significant in this model with *P*-values of 0.012 and 0.000 respectively. **d)** The predicted value for October 2008 is 45,844,197. The predicted value here is smaller than either of the other predictions.

13.81 **a)** $\hat{y}_t = 39.216 + 1.21151y_{t-1} - 0.4338y_{t-2}$. $R^2 = 0.763$ and $s = 0.1853$. This model has a higher R^2 value and a lower standard error estimate. Both lag variables are significant with *P*-values of approximately zero. **b)** The observed number of runs is 42, and the expected number of runs is 45. The *P*-value of the runs test is 0.52. There is no evidence against randomness. **c)** The predicted value for 2008 is 176.017.

Chapter 14: One-Way Analysis of Variance

14.1 **a)** The ANOVA test is for population means, not sample means. **b)** Between-group variance is the variation due to the difference in sample means. **c)** One-way ANOVA is used to compare the means of more than two groups. **d)** Contrasts are used to compare this relation among means.

14.3 $x_{ij} = \mu_i + \varepsilon_{ij}$, $i = 1, 2$, and 3, $j = 1, \ldots, 60$. $I = 3$. The n_i are all equal to 60, and the parameters are μ_1, μ_2, μ_3, and σ.

14.5 **a)** Yes, $130 < 2 \times 85$. **b)** $\hat{\mu}_1 = 75$, $\hat{\mu}_2 = 125$, $\hat{\mu}_3 = 100$, $\hat{\sigma} = 111.58$.

14.7 The histograms below are all skewed to the right. Eyes "down" and "blue" are more strongly skewed than eyes "green" or "brown," which exhibit more of a peak toward the center of the histograms. None of the histograms show Normally distributed values.

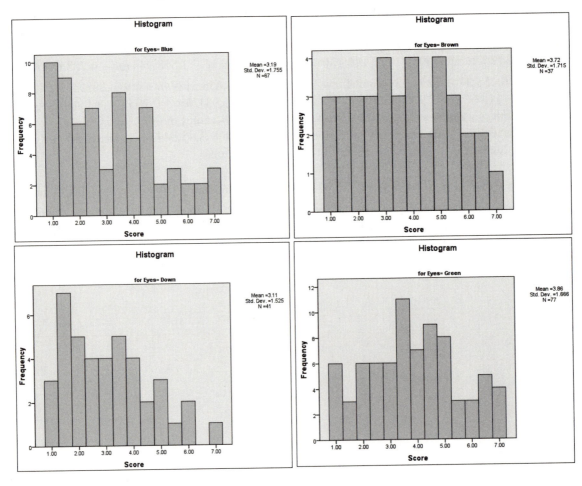

14.9 $24.42 + 613.14 = 637.56$.

14.11 $3 + 218 = 221$.

14.13 MST $= 637.56/221 = 2.885$. The mean of the 222 observations is 3.497, and the variance is 2.885.

14.15 **a)** 2 and 21. **b)** 3.47. **c)** 3.15. **d)** The degrees of freedom in the numerator are the same for both situations, but the degrees of freedom in the in the denominator is much larger with the increased sample size. As the denominator degrees of freedom increases, the critical F value becomes smaller (it is easier to reject the null hypothesis with a larger sample size).

14.17 Applet.

14.19 **a)** SAS labels the standard error as Root MSE because the value can be found by taking the square root of the MSE. **b)** $\sqrt{0.616003} = 0.7849$.

14.21 $SE_C = s_p \sqrt{\dfrac{a_i^2}{n_i}} = 8\sqrt{\dfrac{1}{16}\left[\left(\dfrac{1}{2}\right)^2 + \left(\dfrac{1}{2}\right)^2 + \left(-\dfrac{1}{2}\right)^2 + \left(-\dfrac{1}{2}\right)^2\right]} = 2.$

14.23 (3.02, 10.98) using 80 degrees of freedom on Table D to be conservative.

14.25 $t_{23} = \dfrac{\overline{x}_2 - \overline{x}_3}{s_p\sqrt{\dfrac{1}{n_2} + \dfrac{1}{n_3}}} = \dfrac{46.7273 - 44.2727}{6.31\sqrt{\dfrac{1}{22} + \dfrac{1}{22}}} = 1.29.$

14.27 We would fail to reject the null hypothesis.

14.29 2.455/1.904 = 1.2894.

14.31 Mark groups 1, 2, and 4 with A, groups 2 and 3 with B. Groups 1, 2, and 4 do not differ significantly from each other. Groups 2 and 3 do not differ significantly from each other. However, group 3 does differ significantly from groups 1 and 4.

14.33 (–2.23, 7.14). Yes, the interval includes 0.

14.35 **a)**

Sample Size	10	20	30	40	50	100
Power	0.275	0.532	0.726	0.851	0.923	0.998

b) The power increases as the sample size increases. See the scatterplot below.

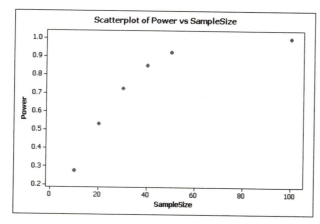

c) Based on the sample sizes selected, a sample of size 50 would be appropriate. The power is fairly high at 0.923. Increasing the sample size to 100 increases the power so it is very close to one, but to get there the sample size must be doubled.

14.37   **a)** $F(4, 30)$.

p	0.100	0.050	0.025	0.010	0.001
F	2.14	2.69	3.25	4.02	6.12

b)

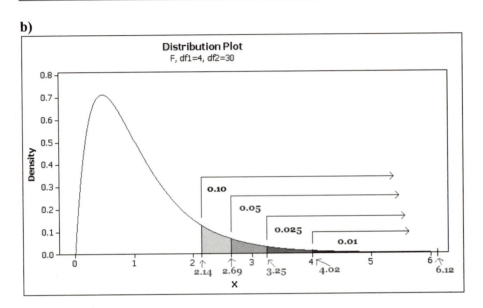

c) $0.025 < P\text{-value} < 0.05$.   **d)** You can never conclude from an F test that all the means are different. The alternative hypothesis is that "not all the means are the same." In this case, we would reject the null hypothesis that all the means are equal.

14.39   **a)** The degrees of freedom are 4 and 40, resulting in an F-statistic of 1.54 with a P-value of 0.2091.

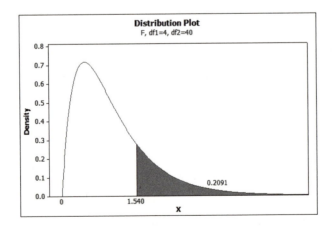

b) The degrees of freedom are 2 and 21, resulting in an F-statistic of 3 with a P-value of 0.0714.

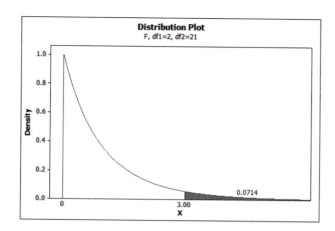

14.41 **a)** Response variable is a 1 to 5 score for the question, "Did you discuss the presentation with any of your friends?" $I = 3$, $n_1 = 220$, $n_2 = 145$, $n_3 = 76$, and $N = 441$. Populations being compared are students in three different groups, elementary statistics students, health and safety students, and students in cooperative housing units. **b)** Response variable is a fee acceptance rating from 1 to 5. $I = 3$, $n_1 = 31$, $n_2 = 18$, $n_3 = 45$, and $N = 94$. Populations are students who do not attend varsity games, students who attend football or basketball varsity games, and students who attend other varsity competitions.
c) Response variable is sales. $I = 4$, $n_i = 5$, and $N = 20$. Populations compared are customers offered a free drink, customers offered a free cookie, customers offered free chips, and customers offered nothing free.

14.43 **a)** $H_0 : \mu_1 = \mu_2 = \mu_3$ and H_a: Not all the means are the same. DFG = 2, DFE = 438, DFT = 440, $F(2, 438)$. **b)** $H_0 : \mu_1 = \mu_2 = \mu_3$ and H_a: Not all the means are the same. DFG = 2, DFE = 91, DFT = 93, $F(2, 91)$. **c)** $H_0 : \mu_1 = \mu_2 = \mu_3 = \mu_4$ and H_a: Not all the means are the same. DFG = 3, DFE = 16, DFT = 19, $F(3, 16)$.

14.45 Answers will vary.

14.47 **a)** No, the biggest s (25) is not less than twice the smallest s (12). **b)** $s_1^2 = 625$, $s_2^2 = 441$, $s_3^2 = 144$, $s_2^2 = 529$. **c)** $s_p^2 = 232.67$. **d)** $s_p = 15.25$. **e)** The third group had the largest sample size, so it had the heaviest weight in the pooled standard deviation.

14.49 **a)** The last question does not result in rejecting the null hypothesis, so there is no evidence of a difference between the group means for this question. **b)** Using Excel, $t^{**} = 2.25$. The following table gives the test statistics. Those that have significant differences are highlighted.

Question	t_{12}	t_{13}	t_{14}	t_{15}	t_{23}	t_{24}	t_{25}	t_{34}	t_{35}	t_{45}
Culture	−0.74	−0.79	−2.89	−0.72	0.03	−2.50	0.10	−2.73	0.08	2.75
Tour	−0.55	2.38	1.37	0.59	3.55	2.00	1.41	−0.51	−2.47	−1.07
Sports	0.27	0.32	−1.15	1.67	0.03	−1.51	1.63	−1.65	1.91	2.86

Those who went to Hawaii for leisure had a significantly different average response to the question about experiencing native Hawaiian culture than the other groups. Those who went to Hawaii for sports had a significantly different average response about preferring a group tour than all other groups except those in Hawaii for leisure purposes.

Finally, those who went to Hawaii for leisure and those who went for business had significantly different responses to the question about wanting to experience ocean sports.

14.51   a) 2 and 117. **b)** The *P*-value from Excel is 0.000298, so we can conclude that at least one of the bargainers received a different average price reduction than the others. **c)** This study only measured results for three people. We don't know if race and gender were the reason for the results or if it had more to do with the individual personalities. Results should not be generalized without further study.

14.53   a) Yes, 2(0.657) > 0.824. $s_p = 0.768$. **b)** The degrees of freedom are 2 and 767. The *P*-value from Excel is 3.175E-08. There is significant evidence that at least one of the groups gave a different average rating for dining satisfaction. The graph of the *F* distribution is below. The *F*-statistic would be far to the right, so there is little chance of this result occurring due to chance if the dining satisfaction was the same for all three groups.

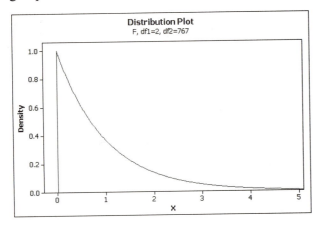

c) Label students with meal plans a group 1, faculty with meal plans as group 2, and students without meal plans as group 3. $\psi = \frac{1}{2}(\mu_1 + \mu_3) - \mu_2$. $H_0: \frac{1}{2}(\mu_1 + \mu_3) = \mu_2, H_a: \frac{1}{2}(\mu_1 + \mu_3) < \mu_2$. The test statistic is $t = -5.98$ with 767 degrees of freedom. The *P*-value is less than 0.0001. There is evidence that the faculty satisfaction rating is higher on average.

14.55   a)

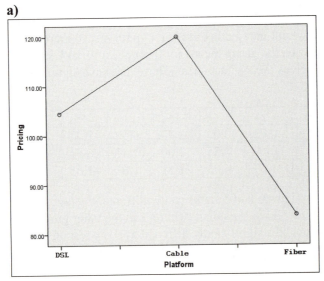

There does appear to be a difference in pricing, but whether that difference is significant can not be determined from a means plot. **b)** Since two times the smallest standard deviation is larger than the largest standard deviation, it is reasonable to pool the variances. **c)** $F(2, 44) = 3.39$. *P*-value = 0.0427. There is evidence that at least one platform has different pricing than the others.

14.57 **a)** 3 and 2286. **b)** $F = 2.5304$. **c)** *P*-value = 0.056. At the 5% significance level, we would not conclude that there is any significant difference in the mean scores across classes; however, the result would be significant at the 10% level.

14.59 **a)** In general, as the number of accommodations increased, the mean grade decreased, but it is not a steady decline. (See the means plot that follows).

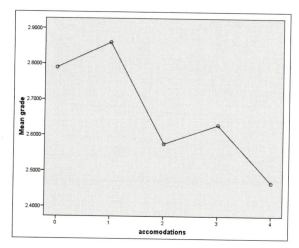

b) So many decimal places are not necessary. No additional information is gained by having so many. Using 2 decimal places would be fine. **c)** The biggest *s* is not exactly less than 2 times the smallest *s* (1.66233 vs. 2 x 0.82745), but it is close. Pooling should not be used (s_p would be 0.86). **d)** It is never a good idea to eliminate data points without a good reason. Since the mean grade is affected similarly for 2, 3, and 4 accommodations, it is not unreasonable to group these data together. **e)** These data do not represent 245 independent observations. Some students were measured multiple times. **f)** Answers will vary. University policies, teacher policies, and student backgrounds might not be the same from school to school. **g)** There is no control group, so it is impossible to comment on the effectiveness of the accommodations.

14.61 **a)** Checking whether the largest $s < 2 \times$ smallest *s* for each group indicates that it is appropriate to pool the standard deviations for intensity and recall but not frequency. **b)** $F(4, 405)$. The *P*-value for all three one-way ANOVA tests is < 0.001, so for frequency, intensity, and recall there is strong evidence that not all the means are the same.

p	0.100	0.050	0.025	0.010	0.001
F	1.95	2.42	2.85	3.41	4.81

c) Hispanic Americans were highest in each of the three groups, Asian and Japanese were both low, and European and Indian were close together in the middle. **d)** Answers will vary. People near a university might not be representative of all citizens of those countries. **e)** The chi-square test has a test statistic of 11.353 and a *P*-value of 0.023, so at the 5% significance level, there is enough evidence to say that there is a relationship between gender and culture. In other words, there is evidence that the proportion of men and women is not the same for all of the cultural groups. This may affect how broadly the results can be applied.

14.63 **a)**

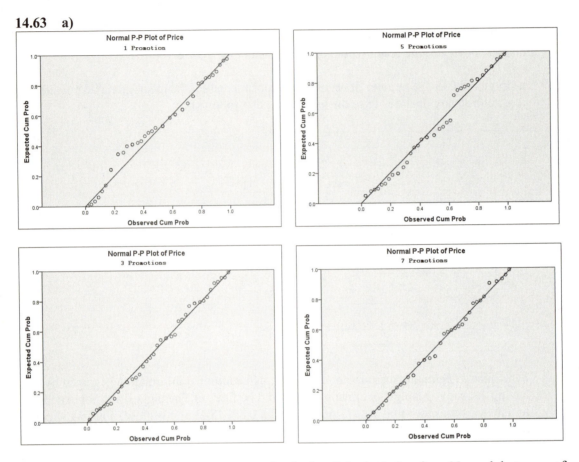

The group that saw only one promotion had a slight deviation from Normal, but none of the distributions are significantly non-Normal.

b)

Number of Promotions	Average	Standard Deviation	Sample Size
1	4.56	0.36	40
3	4.42	0.34	40
5	4.15	0.29	40
7	3.98	0.32	40

c) Yes, the assumption of equal standard deviations is appropriate because $0.36 < 2 \times 0.29$. **d)** H_0: All μs are equal. H_a: At least one μ is different from the others. The test statistic is the F statistic with 3 and 156 degrees of freedom. $F = 25.655$ with a *P*-value of 0.0. There is at least one μ that is significantly different from the others.

14.65 $\psi = \frac{1}{2}(\mu_1 + \mu_2) - \mu_3$. H_0: $\frac{1}{2}(\mu_1 + \mu_2) = \mu_3$, H_a: $\frac{1}{2}(\mu_1 + \mu_2) \neq \mu_3$. $c = -9.695$ and $SE_c = 2.736$. $t = -3.54$ with 587 degrees of freedom and a *P*-value = 0.000432. There is evidence at the 5% level that supervisors have different average SCI scores than workers.

14.67 a)

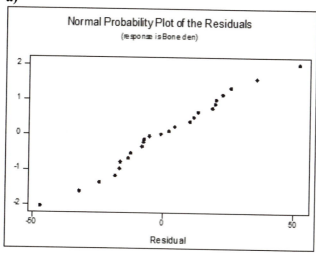

b)

Multiple Comparisons

Dependent Variable: Bone Density

Bonferroni

(I) Treatment	(J) Treatment	Mean Difference (I-J)	Std. Error	Sig.	95% Confidence Interval	
					Lower Bound	Upper Bound
1	2	-11.400	9.653	.744	-36.04	13.24
	3	-37.600*	9.653	.002	-62.24	-12.96
2	1	11.400	9.653	.744	-13.24	36.04
	3	-26.200*	9.653	.034	-50.84	-1.56
3	1	37.600*	9.653	.002	12.96	62.24
	2	26.200*	9.653	.034	1.56	50.84

*. The mean difference is significant at the .05 level.

There is a significant difference in bone density between the high-jump group (group 3) and the control (group 1) and low-jump (group 2) groups. There is no significant difference between the control and low-jump groups.

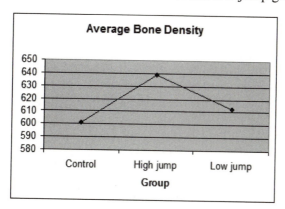

14.69 a)

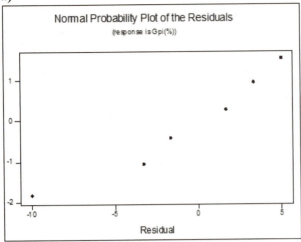

The Normal assumption appears reasonable.

b) Tukey's pairwise comparisons
Family error rate = 0.0500
Individual error rate = 0.00569
Critical value = 4.75

Intervals for (column level mean) – (row level mean)

	ECM1	ECM2	ECM3	MAT1	MAT2
ECM2	–10.43				
	13.76				
ECM3	–20.43	–22.09			
	3.76	2.09			
MAT1	29.57	27.91	37.91		
	53.76	52.09	62.09		
MAT2	46.24	44.57	54.57	4.57	
	70.43	68.76	78.76	28.76	
MAT3	41.24	39.57	49.57	–0.43	–17.09
	65.43	63.76	73.76	23.76	7.09

There is a significant difference between each of the three ECMs and the MATs. There is a significant difference between MAT1 and MAT2. There is no significant difference between the ECMs or between MAT2 and MAT3 or MAT1 and MAT3.

14.71 a)

Source	DF	Sum of Squares	Mean Square	F
Groups	3	104,855.87	34,951.96	15.86
Error	32	70,500.59	2,203.14	
Total	35	175,356.46	5,010.18	

b) H_0: All μ's are equal; H_a: At least one μ is not equal to the rest. **c)** $F(3, 32)$. P-value $= 0.000$. At least one μ is different from the others. **d)** $s_p^2 = 2,203.14$, $s_p = 46.94$.

14.73   **a)** $s_p^2 = 3.8975$, MSE. **b)** See chart below.

Source	DF	Sum of Squares	Mean Square	F
Groups	2	17.22	8.61	2.2091
Error	206	802.89	3.90	
Total	208			

c) H_0: All μ's are equal and H_a: At least one μ is not equal to the rest. **d)** $F(2, 206)$. P-value $= 0.1124$. There is no strong evidence to support the alternative hypothesis.

14.75   **a)** $\psi = \mu_{brown} - \frac{1}{2}(\mu_{blue} + \mu_{green})$. **b)** $\psi = \mu_{down} - \frac{1}{3}(\mu_{blue} + \mu_{green} + \mu_{brown})$.

14.77   **a)** For contrast ψ_1, test H_0: $\psi_1 = 0$ vs. H_a: $\psi_1 > 0$. For the contrast ψ_2, test H_0: $\psi_2 = 0$ vs. H_a: $\psi_2 \neq 0$. **b)** $c_1 = 0.195$. $c_2 = -0.48$. **c)** $SE_{c1} = 0.310$, $SE_{c2} = 0.293$. **d)** The test statistic for H_0: $\psi_1 = 0$, H_a: $\psi_1 > 0$, is $t = 0.629$. P-value > 0.25. There is not enough evidence that the responses were greater for the models with brown eyes than the average scores for those with blue or green eyes. For H_0: $\psi_2 = 0$, H_a: $\psi_2 \neq 0$, the test statistic is $t = -1.636$. $0.05 < P$-value < 0.10. There is not strong evidence that the average response for models with their eyes looking downward is different from the mean of the responses for those whose eyes were visible. **e)** Using 100 df to be more conservative, a 95% confidence interval for ψ_1 is 0.195 ± 0.615, and a 95% confidence interval for ψ_2 is -0.48 ± 0.582.

14.79   **a)** $\psi_1 = \mu_T - \mu_C$. H_{01}: $\mu_T = \mu_C$, H_{a1}: $\mu_T > \mu_C$. $\psi_2 = \mu_T - \frac{1}{2}(\mu_C + \mu_S)$. H_{02}: $\mu_T = \frac{1}{2}(\mu_C + \mu_S)$, H_{a2}: $\mu_T > \frac{1}{2}(\mu_C + \mu_S)$. $\psi_3 = \mu_J - \frac{1}{3}(\mu_T + \mu_C + \mu_S)$. H_{03}: $\mu_J = \frac{1}{3}(\mu_T + \mu_C + \mu_S)$, H_{a3}: $\mu_J > \frac{1}{3}(\mu_T + \mu_C + \mu_S)$. **b)** $t_1 = -0.6635$ with a P-value $= 0.2559$. $t_2 = 1.242$ with a P-value $= 0.1116$. $t_3 = 5.282$ with a P-value $= 0.000$. The average score for the joggers groups is significantly higher than the average of the other three groups. **c)** This study does not show causation. There are many lurking variables that could contribute to a low score.

14.81

Group (I)	Group (J)	Mean Difference	Standard Error	t_{ij}
T	C	−17.06	25.7101	−0.66355
	J	−74.96	20.5096	−3.65488
	S	65.84	20.9922	3.13640
C	T	17.06	25.7101	0.66355
	J	−57.90	25.3176	−2.28695
	S	82.90	25.7101	3.22441
J	T	74.96	20.5096	3.65488
	C	57.90	25.3176	2.28695
	S	140.80	20.5096	6.86509
S	T	−65.84	20.9922	−3.13640
	C	−82.90	25.7101	−3.22441
	J	−140.80	20.5096	−6.86509

The sedentary group is significantly different from the three other groups. The treatment group is significantly different from the jogger group and the sedentary group. There is no significant difference between the jogger group and the control group or between the treatment group and the control group.

14.83

n	DFG	DFE	F^*	λ	Power
50	2	147	3.0576	2.78	0.2950
100	2	297	3.0261	5.56	0.5453
150	2	447	3.0158	8.34	0.7336
175	2	522	3.0130	9.73	0.8017
200	2	597	3.0108	11.12	0.8548

A sample size of 175 gives reasonable power. The gain in power by using 200 women per group may not be worthwhile unless it is easy to get women for the study. If it is difficult or expensive to include more women in the study, one might consider a sample size of 150 per group.

14.85 **a)** Answers will vary. **b)** The power would be 0.641 for a sample size of 50, 0.836 for a sample size of 75, 0.935 for a sample size of 100 and 0.999 for a sample size of 200. A sample size of 100 would be sufficient for this study. **c)** Answers will vary.

14.87 a)

Level	Sample Size	Mean	Standard Deviation
ECM1	3	0.6500	0.0866
ECM2	3	0.6333	0.0289
ECM3	3	0.7333	0.0289
MAT1	3	0.2333	0.0289
MAT2	3	0.6667	0.0289
MAT3	3	0.1167	0.0289

b) These values could have been computed by dividing the means and standard deviations by 100 as well. **c)** $F = 137.94$ with 5 and 12 degrees of freedom. *P*-value = 0.000. These results are identical those from Exercise 14.68.

14.89 **a)** $F(5, 12) = 9.65$ with a *P*-value of 0.001. **b)** Changing this one value decreases the test statistic dramatically and increases the *P*-value. Outliers can have a substantial impact on the results of an ANOVA. **c)** See chart below.

Level	Sample Size	Mean	Standard Dev.
ECM1	3	48.33	37.53
ECM2	3	63.33	2.89
ECM3	3	73.33	2.89
MAT1	3	23.33	2.89
MAT2	3	6.67	2.89
MAT3	3	11.67	2.89

The standard deviation for the ECM1 group is very large compared to the standard deviations of the other groups.

14.91 a)

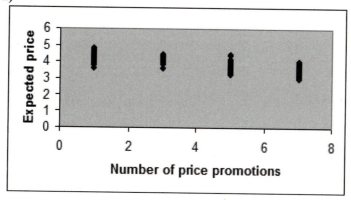

The pattern is roughly linear. **b)** The test for $\beta = 0$ is the test for no linear relationship between explanatory variable and response variable. **c)** The results of the regression analysis show that there is a significant linear relationship between the number of promotions and the expected price. The ANOVA results state that there is at least one μ that is different from the other three. Regression analysis gives more information because we know that there is a linear relationship. ANOVA does not tell us the type of relationship between the variables.

Case Study 14.1

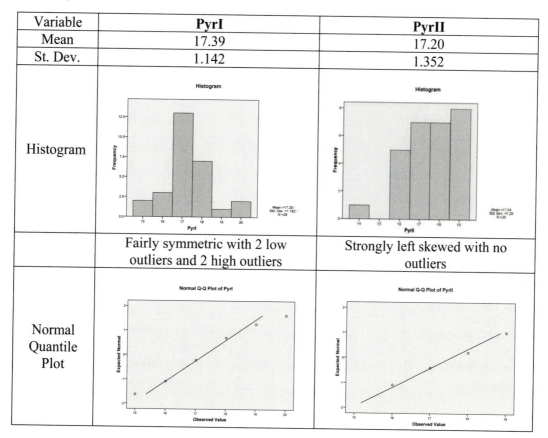

Variable	PyrI	PyrII
Mean	17.39	17.20
St. Dev.	1.142	1.352
Histogram		
	Fairly symmetric with 2 low outliers and 2 high outliers	Strongly left skewed with no outliers
Normal Quantile Plot		

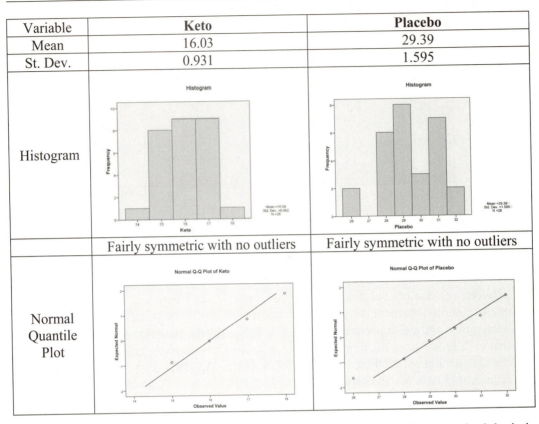

Variable	Keto	Placebo
Mean	16.03	29.39
St. Dev.	0.931	1.595
Histogram	Fairly symmetric with no outliers	Fairly symmetric with no outliers
Normal Quantile Plot		

The largest standard deviation (1.595) is less than twice the smallest standard deviation (0.962), so it is appropriate to pool the standard deviations. The PyrI group does have come outliers, which is not ideal, and the sample size for the Placebo group is much smaller than the sample size for the other groups. (The numbers above are from SPSS. Minitab gives slightly different results.)

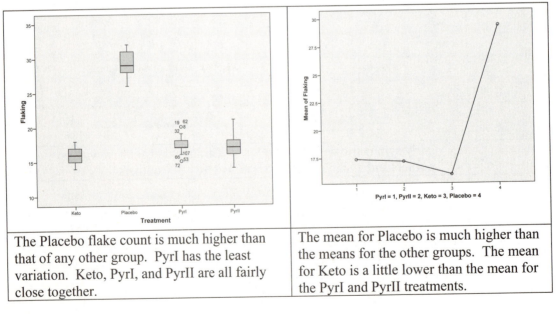

| The Placebo flake count is much higher than that of any other group. PyrI has the least variation. Keto, PyrI, and PyrII are all fairly close together. | The mean for Placebo is much higher than the means for the other groups. The mean for Keto is a little lower than the mean for the PyrI and PyrII treatments. |

The ANOVA test has hypotheses $H_0 : \mu_{PyrI} = \mu_{PyrII} = \mu_{Keto} = \mu_{Placebo}$ and H_a : Not all of the means are the same. The F test statistic is 967.819, and the P-value is close to 0. Therefore, we have strong evidence that not all the means are the same (at least one mean is different).

ANOVA

Flaking

	Sum of Squares	df	Mean Square	F	Sig.
Between Groups	4151.428	3	1383.809	967.819	.000
Within Groups	501.868	351	1.430		
Total	4653.296	354			

Using the Bonferroni multiple comparisons test, PyrI is not significantly different from PyrII, but Keto is significantly different from all three other groups, and Placebo is significantly different from all three other groups.

Multiple Comparisons

Dependent Variable: Flaking

Bonferroni

(I) PyrI = 1, PyrII = 2, Keto = 3, Placebo = 4	(J) PyrI = 1, PyrII = 2, Keto = 3, Placebo = 4	Mean Difference (I-J)	Std. Error	Sig.	95% Confidence Interval	
					Lower Bound	Upper Bound
1	2	.191	.161	1.000	-.24	.62
	3	1.365*	.162	.000	.93	1.79
	4	-12.000*	.253	.000	-12.67	-11.33
2	1	-.191	.161	1.000	-.62	.24
	3	1.174*	.163	.000	.74	1.61
	4	-12.191*	.253	.000	-12.86	-11.52
3	1	-1.365*	.162	.000	-1.79	-.93
	2	-1.174*	.163	.000	-1.61	-.74
	4	-13.365*	.254	.000	-14.04	-12.69
4	1	12.000*	.253	.000	11.33	12.67
	2	12.191*	.253	.000	11.52	12.86
	3	13.365*	.254	.000	12.69	14.04

*. The mean difference is significant at the .05 level.

Contrast #1: $\psi_1 : \frac{1}{3}(\mu_{PyrI} + \mu_{PyrII} + \mu_{Keto}) - \mu_{Placebo}$ with $H_0 : \psi_1 = 0$ and $H_a : \psi_1 < 0$

(the two-sided hypothesis would also be appropriate, but the researchers probably expect that the treatments will reduce flaking over the Placebo group so a one-sided test is chosen here). The value of the contrast is $c = -12.39$, $SE_c = 0.233$, $t = -53.167$, df $= 351$, and P-value close to 0 for either the one-sided or two-sided test. Therefore, we have evidence that the average flaking count of the three treatment groups is significantly lower than the average of the Placebo group.

Contrast #2: $\psi_2 : \frac{1}{2}(\mu_{PyrI} + \mu_{PyrII}) - \mu_{Keto}$ with $H_0 : \psi_2 = 0$ and $H_a : \psi_2 \neq 0$. The value of the contrast is $c = 1.27$, $SE_c = 0.141$, $t = 8.983$, df $= 351$, and P-value close to 0 for either the one-sided or two-sided test. Therefore, we have evidence that the average flaking count of the three treatment groups is significantly different (or higher of if one-sided test is done using a ">" in the alternative hypothesis) in the Keto group than in the average of the PyrI and PyrII groups.

Contrast #3: $\psi_3 : \mu_{PyrI} - \mu_{PyrII}$ with $H_0 : \psi_3 = 0$ and $H_a : \psi_3 \neq 0$. The value of the contrast is $c = 0.19$, $SE_c = 0.161$, $t = 1.187$, df $= 351$, and P-value is 0.236 for two-sided test. Therefore, we do not have enough evidence to say that the average flaking count is significantly different for the PyrI and PyrII groups.

Contrast Coefficients

| | PyrI = 1, PyrII = 2, Keto = 3, Placebo = 4 | | | |
Contrast	1	2	3	4
1	.33	.33	.33	-.99
2	.5	.5	-1	0
3	1	-1	0	0

Contrast Tests

		Contrast	Value of Contrast	Std. Error	t	df	Sig. (2-tailed)
Flaking	Assume equal variances	1	-12.39	.233	-53.167	351	.000
		2	1.27	.141	8.983	351	.000
		3	.19	.161	1.187	351	.236
	Does not assume equal variances	1	-12.39	.305	-40.635	29.465	.000
		2	1.27	.124	10.268	266.704	.000
		3	.19	.169	1.133	211.008	.258

Case Study 14.2

In the SPSS output below, group B = 1, group D = 2, and group S = 3. For each of these tests, it is appropriate to pool the standard deviations because the biggest s for each test is smaller than twice the smallest s.

Descriptives

		N	Mean	Std. Deviation	Std. Error	95% Confidence Interval for Mean		Minimum	Maximum
						Lower Bound	Upper Bound		
Pre1	1	22	10.50	2.972	.634	9.18	11.82	4	16
	2	22	9.73	2.694	.574	8.53	10.92	6	16
	3	22	9.14	3.342	.713	7.65	10.62	4	14
	Total	66	9.79	3.021	.372	9.05	10.53	4	16
Pre2	1	22	5.27	2.763	.589	4.05	6.50	2	13
	2	22	5.09	1.998	.426	4.21	5.98	1	8
	3	22	4.95	1.864	.397	4.13	5.78	2	9
	Total	66	5.11	2.213	.272	4.56	5.65	1	13
Post1	1	22	6.68	2.767	.590	5.46	7.91	2	12
	2	22	9.77	2.724	.581	8.56	10.98	5	14
	3	22	7.77	3.927	.837	6.03	9.51	1	15
	Total	66	8.08	3.394	.418	7.24	8.91	1	15
Post2	1	22	5.55	2.041	.435	4.64	6.45	3	10
	2	22	6.23	2.092	.446	5.30	7.15	0	11
	3	22	8.36	2.904	.619	7.08	9.65	1	13
	Total	66	6.71	2.636	.324	6.06	7.36	0	13
Post3	1	22	41.05	5.636	1.202	38.55	43.54	32	54
	2	22	46.73	7.388	1.575	43.45	50.00	30	57
	3	22	44.27	5.767	1.229	41.72	46.83	33	53
	Total	66	44.02	6.644	.818	42.38	45.65	30	57

The ANOVA F tests results follow.

Neither of the pretests give evidence for significant differences between the groups. It does seem that the three groups of subjects were similar at the start of the study.

All three post-tests show evidence that not all the means are the same within each test (at least one group has a different mean within each test).

ANOVA

		Sum of Squares	df	Mean Square	F	Sig.
Pre1	Between Groups	20.576	2	10.288	1.132	.329
	Within Groups	572.455	63	9.087		
	Total	593.030	65			
Pre2	Between Groups	1.121	2	.561	.111	.895
	Within Groups	317.136	63	5.034		
	Total	318.258	65			
Post1	Between Groups	108.121	2	54.061	5.317	.007
	Within Groups	640.500	63	10.167		
	Total	748.621	65			
Post2	Between Groups	95.121	2	47.561	8.407	.001
	Within Groups	356.409	63	5.657		
	Total	451.530	65			
Post3	Between Groups	357.303	2	178.652	4.481	.015
	Within Groups	2511.682	63	39.868		
	Total	2868.985	65			

The results of the Bonferroni multiple comparison tests are below. Ignore the results for both pretests since there are no significant differences to be found there.

Post1: There is a significant difference between group B and group D, but there are no other significant differences.

Post2: Group S is significantly different from both group B and group D, but group B and group D are not significantly different from each other.

Post3: There is a significant difference between group B and group D, but there are no other significant differences.

Post-test 1 and post-test 3 indicate that teaching method D is more effective than the others.

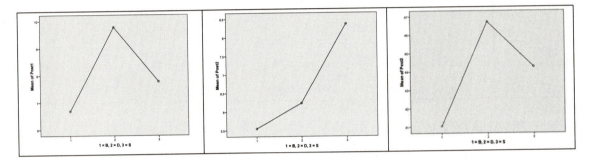

Multiple Comparisons

Bonferroni

Dependent Variable	(I) 1 = B, 2 = D, 3 = S	(J) 1 = B, 2 = D, 3 = S	Mean Difference (I-J)	Std. Error	Sig.	95% Confidence Interval	
						Lower Bound	Upper Bound
Pre1	1	2	.773	.909	1.000	-1.46	3.01
		3	1.364	.909	.416	-.87	3.60
	2	1	-.773	.909	1.000	-3.01	1.46
		3	.591	.909	1.000	-1.64	2.83
	3	1	-1.364	.909	.416	-3.60	.87
		2	-.591	.909	1.000	-2.83	1.64
Pre2	1	2	.182	.676	1.000	-1.48	1.85
		3	.318	.676	1.000	-1.35	1.98
	2	1	-.182	.676	1.000	-1.85	1.48
		3	.136	.676	1.000	-1.53	1.80
	3	1	-.318	.676	1.000	-1.98	1.35
		2	-.136	.676	1.000	-1.80	1.53
Post1	1	2	-3.091*	.961	.006	-5.46	-.73
		3	-1.091	.961	.782	-3.46	1.27
	2	1	3.091*	.961	.006	.73	5.46
		3	2.000	.961	.125	-.36	4.36
	3	1	1.091	.961	.782	-1.27	3.46
		2	-2.000	.961	.125	-4.36	.36
Post2	1	2	-.682	.717	1.000	-2.45	1.08
		3	-2.818*	.717	.001	-4.58	-1.05
	2	1	.682	.717	1.000	-1.08	2.45
		3	-2.136*	.717	.012	-3.90	-.37
	3	1	2.818*	.717	.001	1.05	4.58
		2	2.136*	.717	.012	.37	3.90
Post3	1	2	-5.682*	1.904	.012	-10.36	-1.00
		3	-3.227	1.904	.285	-7.91	1.46
	2	1	5.682*	1.904	.012	1.00	10.36
		3	2.455	1.904	.606	-2.23	7.14
	3	1	3.227	1.904	.285	-1.46	7.91
		2	-2.455	1.904	.606	-7.14	2.23

*. The mean difference is significant at the .05 level.

Chapter 15: Two-Way Analysis of Variance

15.1 **a)** A two-way ANOVA is used when there are two factors and one response variable. **b)** In a 2 × 3 ANOVA, there are two levels for Factor A and 3 levels for Factor B. Each level of Factor A would appear with each (three) levels of Factor B. **c)** The FIT part of the model in a two-way ANOVA consists of the population means μ_{ij}. **d)** You can perform a two-way ANOVA with different sample sizes for all cells, but we assume that all populations have the same standard deviation.

15.3 Response variable: Effectiveness rating. Factors: Training program and Delivery method. $I = 4$, $J = 2$, and $N = 80$.

15.5 The largest difference in per capita income between whites and Asians occurs in the West. The smallest difference is in the Midwest where Asians have a slightly larger per capita income than whites.

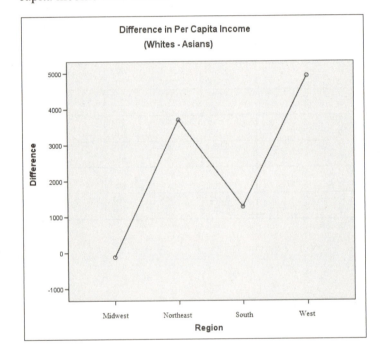

15.7

a) Yes, there is an interaction. The means increase more quickly for level 2 of Factor B than level 1 as Factor A increases.

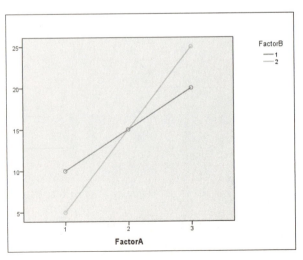

b) No, there is no interaction effect. The lines are parallel.

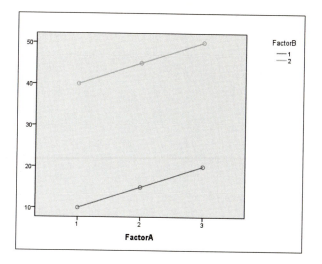

c) Yes, there is an interaction. The lines are not parallel. The means increase for level 1 of Factor B but decrease for level 2 of Factor B as Factor A increases.

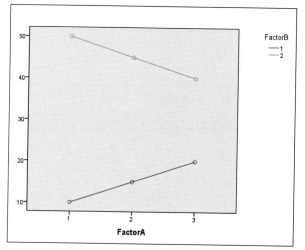

d) Yes, there is an interaction. The effect is not apparent until we reach level 3 of Factor A.

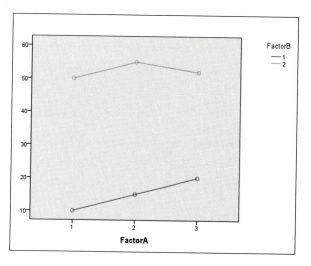

15.9 The effect of the packaging: 2 and 348 degrees of freedom. The effect of the colors: 3 and 348 degrees of freedom. The interaction effect: 6 and 348 degrees of freedom.

15.11

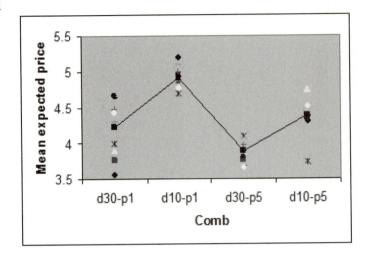

The spread in each group is similar except for the combination d10-p5. There may be an outlier on the low side. The treatment means appear to be different.

15.13

Promo	Discount	Mean	Standard Deviation	N
1	30%	4.225	0.385609	10
1	10%	4.920	0.152023	10
	Total	4.573	0.456611	20
5	30%	3.890	0.162891	10
5	10%	4.393	0.268537	10
	Total	4.142	0.336613	20
Total	30%	4.058	0.335463	20
	10%	4.657	0.343791	20
	Total	4.357	0.452113	40

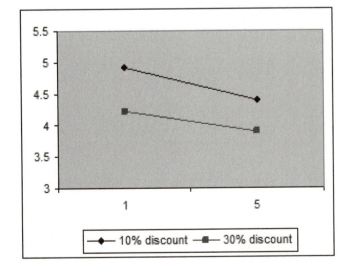

Based on the standard deviations, it makes sense to pool the group standard deviations to get MSE. There is a difference in mean expected price as the number of promotions decrease. There may be an interaction effect due to percentage discount but it is not readily apparent from the plot.

15.15 Answers will vary.

15.17 **a)** The degrees of freedom for interaction will be 2 and 24. The corresponding entries from Table E are shown in the table below.

P-value	*F*
0.100	2.54
0.050	3.40
0.025	4.32
0.010	5.61
0.001	9.34

b)

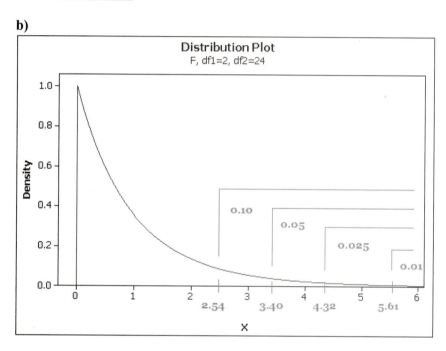

c) 0.0645. **d)** There is no evidence of an interaction at the 5% level, but there is at the 10% level. We would expect the lines to be somewhat, but not exactly, parallel.

15.19 **a)** Response variable: whether the child contracted malaria or not. Factors: initial treatment and daily treatment. $I = 2$, $J = 2$, and $N = 453$. **b)** Response variable: percent of seeds germinating. Factors: number of weeks after harvest and amount of water used in the process. $I = 5$, $J = 2$, and $N = 30$. **c)** Response variable: strength of concrete. Factors: concrete formula and number of freezing cycles. $I = 6$, $J = 3$, and $N = 54$.

15.21 Main effect A (df = 1 and 24) has *P*-value > 0.100. Main effect B (df = 2 and 24) has *P*-value between 0.025 and 0.05. The interaction of A and B (df = 2 and 24) has *P*-value > 0.100. Only the main effect B is significant, and this is significant at both the 5% and 10% levels.

15.23   **a)** Since the lines of the marginal means plot below do intersect, there is an indication of an interaction. Short transaction histories have lower repurchase intent when there is no "thank you" than when there is a "thank you." Long transaction histories result in similar measures of repurchase intent with the intent to repurchase being slightly higher when no "thank you" is given.

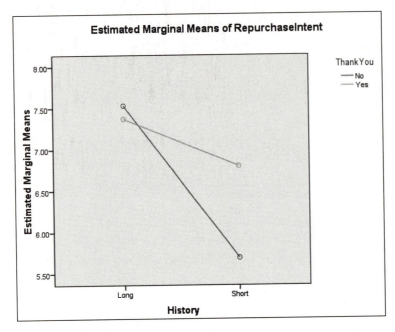

b) The marginal means are 7.45 for long history, 6.245 for short history, 6.61 for no thank you, and 7.085 for thank you. Repurchase intent is generally lower for short histories or for no statements of thanks, but the interaction of these gives much more information.

15.25   **a)** See the marginal means plot that follows. The results for favorable processes are higher than for unfavorable processes for either group of outcomes. Favorable outcomes have higher means than unfavorable outcomes regardless of the process. The lines are not parallel, so there is probably an interaction. There is a bigger difference in the results for the two favorable outcome groups than for the unfavorable outcome groups.

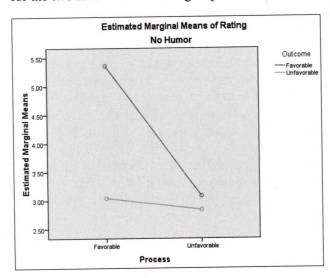

b) The lines for the treatments that used humor are not far from parallel, so there is probably not an interaction. Favorable processes have higher means than unfavorable processes, and favorable outcomes have higher means than unfavorable outcomes.

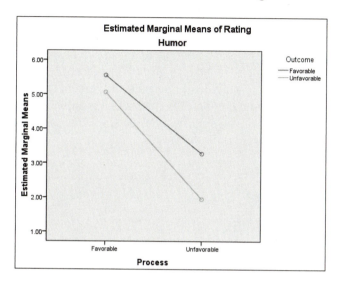

c) Since the results from the two plots are different, there is an indication of a three factor interaction.

15.27 The following tables give the marginal means for all main interactions.

Humor	Mean
Yes	3.88
No	3.6

Process	Mean
Favorable	4.8
Unfavorable	2.74

Outcome	Mean
Favorable	4.28
Unfavorable	3.21

Humor results in greater ratings than no humor, favorable process results in higher ratings than unfavorable process, and favorable outcome results in higher ratings than unfavorable outcome.
The following table shows the marginal means for all two-factor interactions.

Interaction	Mean
No Humor*Favorable Process	4.24
No Humor*Unfavorable Process	2.97
Humor*Favorable Process	5.30
Humor*Unfavorable Process	2.55
No Humor*Favorable Outcome	4.18
No Humor*Unfavorable Outcome	2.94
Humor*Favorable Outcome	4.41
Humor*Unfavorable Outcome	3.41
Favorable Outcome*Favorable Process	5.46
Favorable Outcome*Unfavorable Process	3.17
Unfavorable Outcome*Favorable Process	4.14
Unfavorable Outcome *Unfavorable Process	2.32

The process ratings have a greater difference when humor is used than when no humor is used. The outcome ratings have a greater difference when humor is used than when no humor is used. The outcome ratings have a greater difference when the process rating is favorable than when it is not.

15.29   **a)** S_p = 19.51 with 105 degrees of freedom. **b)** Yes, twice the smallest standard deviation (17.5) is larger than the largest standard deviation (22.1). **c)** Marginal means are 25.1 for individual sender, 19.09 for group sender, 24.27 for individual responder, and 20.74 for group responder. **d)** As seen in the means plot below, there appears to be an interaction since the lines are not parallel. When the sender was a group, the percent returned to them was greater if the responder was an individual; however, the percent returned to an individual was the same regardless of whether the responder was a group or an individual. Since we know that there were no significant effects, we can conclude that the differences seen in the means plot were not substantial.

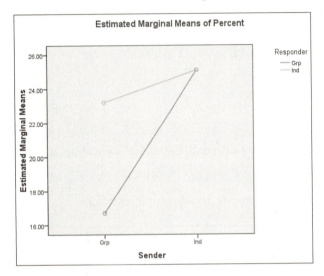

15.31   **a)** The marginal mean for intervention is 11.600, for control 9.967, for baseline 10.000, for 3 months 11.200, and for 6 months 11.150. **b)** See the means plot below. There does appear to be an interaction because the lines are not parallel. The means for the intervention group are higher than the means for the means for the control group in all the different time periods. For the control group (bottom line), the average number of behaviors increases with time. For the intervention group (top line), the average number of behaviors peaks at 3 months from a low of 0 months and then reduces slightly at 6 months.

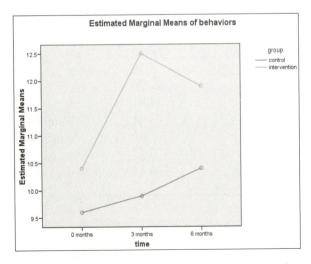

15.33 **a)** The means plot below shows that "familiar" has a higher rating than "unfamiliar" for 1 and 3 repetitions, but "unfamiliar" has a higher rating than "familiar" for 2 repetitions. **b)** Yes. Since the lines intersect, the means plot does suggest an interaction. The pattern for "familiar" is always increasing (although more steeply between 2 and 3 repetitions than between 1 and 2). The pattern for "unfamiliar" shows a much larger mean for 2 repetitions than for 1 and 3.

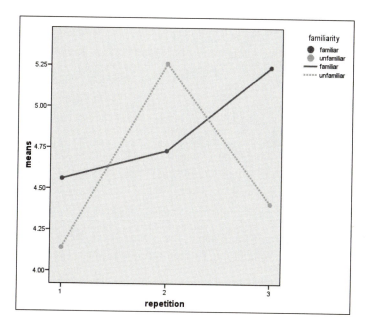

15.35 The pooled estimate of the standard deviation for these data is 1.3077. Since the biggest standard deviation (1.46) is less than twice the smallest standard deviation (1.16), it is reasonable to use a pooled standard deviation for the analysis of these data.

15.37 Subjects were 94 adult staff members at a West Coast university: We do not know the ages or any other details about these subjects. If faculty is included, then these subjects are probably more highly educated than the general public and may watch less television or be less susceptible to the ads. However, staff members at a university may also have a higher income than other segments of the population and may be more familiar with brand names. West Coast residents may have different opinions than residents of other parts of the country. Ninety-four is a fairly large sample size, though.

We do not know what was discussed on the half-hour local news show. Choosing an out-of-state news program was good because the material covered during the show would be fairly neutral to all subjects involved. It would be good to know exactly what topics were covered during the program though and what the subjects' reactions were to the show.

The quality of the ads was prejudged by experts to be "good" and by a sample to be "real," so the quality difference in familiar brands versus unfamiliar brands is probably not an issue, but it would be good to know how those judgments were made. A professional video editor was used to change the brand names, so that work was probably high quality as well.

Most aspects of this research were well done. I would question how well the results will apply to the general public mainly because of how the subjects were selected. It would be good to compare these results to other areas of the country and other types of people.

15.39   a)

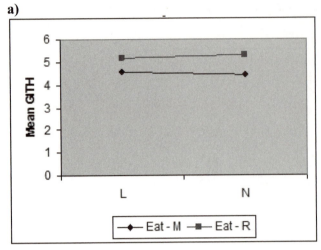

b) The amount of chromium does not seem to affect the mean GITH. Restricting the diet does seem to increase the amount of GITH. No interaction is obvious.

c)

Chromium	Eat - M	Eat - R	Means Difference
L	4.545	5.175	0.630
N	4.425	5.317	0.892
Total means	4.485	5.246	0.761

The differences in means are similar. No interaction is obvious.

15.41   Excel output:

ANOVA: Two-Factor with Replication

SUMMARY	4 weeks	8 weeks	Total
ECM1			
Count	3	3	6
Sum	195	190	385
Average	65	63.33333	64.16667
Variance	75	8.333333	34.16667
ECM2			
Count	3	3	6
Sum	190	190	380
Average	63.33333	63.33333	63.33333
Variance	8.333333	33.33333	16.66667

ECM3			
Count	3	3	6
Sum	220	220	440
Average	73.33333	73.33333	73.33333
Variance	8.333333	33.33333	16.66667

MAT1			
Count	3	3	6
Sum	70	65	135
Average	23.33333	21.66667	22.5
Variance	8.333333	33.33333	17.5

MAT2			
Count	3	3	6
Sum	20	20	40
Average	6.666667	6.666667	6.666667
Variance	8.333333	8.333333	6.666667

MAT3			
Count	3	3	6
Sum	35	30	65
Average	11.66667	10	10.83333
Variance	8.333333	25	14.16667

b)

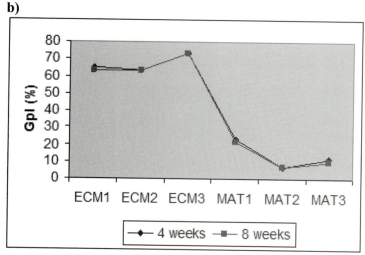

The difference between the ECM and MAT is dramatic. There does not seem to be a difference in mean Gpi between the two times. **c)** For the scaffold material effect: $F = 251.26$ with df = 5 and 24. $P = 0.0$. For the time effect: $F = 0.29$ with df = 1 and 24. $P = 0.595$. For the interaction effect: $F = 0.058$ with df = 5 and 24. $P = 0.998$. The main effect of type of material used as a scaffold was significant. The effect of time and the interaction effect showed no significant differences in means.

15.43 For the 2-week period: There is a significant difference (using Tukey's comparison with alpha = 0.05) between all three MAT groups and between the MAT groups and each of the ECM groups. There is no significant difference between the three ECM groups. For the 4-week period: There is a significant difference (using Tukey's comparison with alpha = 0.05) between the three MAT groups and the three ECM groups. There is no significant difference among the MAT groups or the ECM groups. For the 8-week period: The same result holds as for the 4-week period.

15.45   Yes, in general, this data support the hypothesis that foods cooked in iron pots contain a significantly higher iron content than foods cooked in aluminum or clay pots. The interaction effect is small compared to the main effect of type of pot.

15.47 a)

Tool	Time	Mean	Standard Deviation
1	1	25.0307	0.001155
	2	25.0280	0
	3	25.0260	0
2	1	25.0167	0.001155
	2	25.0200	0.002000
	3	25.0160	0
3	1	25.0063	0.001528
	2	25.0127	0.001155
	3	25.0093	0.001155
4	1	25.0120	0
	2	25.0193	0.001155
	3	25.0140	0.004000
5	1	24.9973	0.001155
	2	25.0060	0
	3	25.0003	0.001528

b)

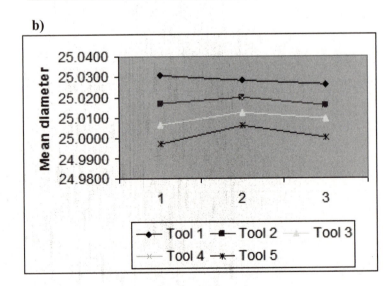

The means change with each tool and with time but the times the measurements were taken does not appear as dramatic. There may be a slight interaction effect.

c)

ANOVA

Source of Variation	SS	df	MS	F	P-value	F crit
Sample	0.003597	4	0.000899	412.9439	9.27E-26	2.689632
Columns	0.00019	2	9.5E-05	43.60204	1.33E-09	3.315833
Interaction	0.000133	8	1.67E-05	7.645409	1.55E-05	2.266162
Within	6.53E-05	30	2.18E-06			
Total	0.003986	44				

Both the main effects of time and tool type were significant, as well as the interaction effect.

15.49 a)

Number of Promotions	Percent Discount	Mean	Standard Deviation
1	40	4.423	0.184755
	30	4.225	0.385609
	20	4.689	0.233069
	10	4.920	0.152023
3	40	4.284	0.204026
	30	4.097	0.234618
	20	4.524	0.270727
	10	4.756	0.242908
5	40	4.058	0.175992
	30	3.890	0.162891
	20	4.251	0.264846
	10	4.393	0.268537
7	40	3.780	0.214372
	30	3.760	0.261789
	20	4.094	0.240749
	10	4.269	0.269916

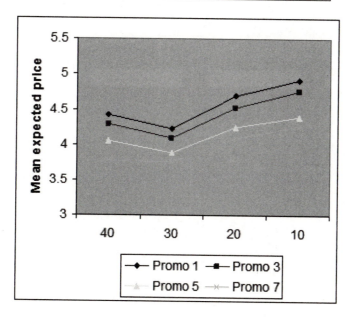

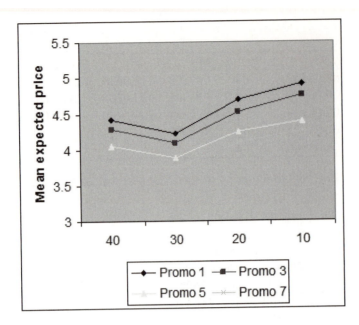

There is a main effect from the number of promotions offered. As the number of promotions offered increases, the expected price decreases. There is also a main effect from the percent discount. As the percent discount increases, the expected price tends to decrease. There does not appear to be an interaction effect.

b)

ANOVA

Source of Variation	SS	df	MS	F	P-value	F crit
Sample	8.360502	3	2.786834	47.72504	1.78E-21	2.667441
Columns	8.306937	3	2.768979	47.41927	2.24E-21	2.667441
Interaction	0.230586	9	0.025621	0.438758	0.9121	1.945452
Within	8.40867	144	0.058394			
Total	25.30669	159				

The main effects of promotions offered ($F = 47.72$ with df = 3 and 144 and $P = 0.0$) and percent discount ($F = 47.42$ with df = 3 and 144 and $P = 0.0$) are both statistically significant. The interaction effect is not statistically significant ($F = 0.4388$ with df = 9 and 144 and $P = 0.9121$).

15.51 a)

ANOVA

Source of Variation	df	SS	MS	F
A (Chromium)	1	0.00121	0.00121	0.04031
B (Eat)	1	5.79121	5.79121	192.91173
AB	1	0.17161	0.17161	5.71652
Error	36	1.08084	0.03002	
Total	39	7.04487		

b) $F = 5.71652$. The distribution has df $= 1$ and 36. $P = 0.022125$. **c)** Chromium: $F = 0.04031$ with df $= 1$ and 36. $P = 0.842006$. Eat: $F = 192.91173$ with df $= 1$ and 36. $P = 0.0$. **d)** $s_p^2 = 0.03002$ and $s_p = 0.17326$. **e)** The results of this ANOVA test are consistent with the observations made in Exercise 15.39. The chromium level does not affect the level of GITH. There is a significant difference in GITH between the restricted diet and unrestricted diet. The interaction between chromium and eat is statistically significant at the 0.05 level, but not at the 0.01 level. The interaction effect is relatively small compared to the main effect of eat.

15.53 **a)** $F = 22.36$ with df $= 1$ and 945: $P = 0.0$. $F = 37.44$ with df $= 1$ and 945: $P = 0.0$. $F = 2.10$ with df $= 1$ and 945: $P = 0.1476$. **b)** The main effects of gender and handedness are significant but the interaction effect is not.

15.55 **a)** The means plot below shows females with a higher fear of being called a nerd or a teacher's pet than males in all three countries. The difference between genders was much smaller in Canada than either Germany or Israel. Since only the interaction has a P-value less than 0.05, we can conclude that the differences seen in country and gender are not significant but their interaction is. **b)** While Germany's results are technically lower than Canada's and Israel's, the results are not significant; therefore, the fact that the ratings are lower could just be due to random chance. **c)** Students who have higher grades in mathematics have the reputation for being more likely to be called a nerd or a teacher's pet as well as valuing high achievement. So taking this variable into account might be beneficial in helping to address this hypothesis.

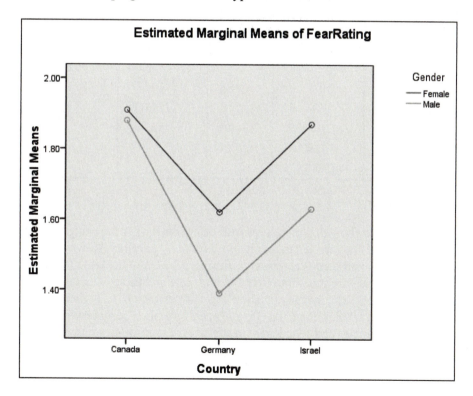

Case Study 15.1

The means for the combinations of factors are shown in the table below. The means plot is shown as well.

	Male	Female
Control	130.00	148.00
Runners	103.98	115.99

The results of a two-way ANOVA analysis using SPSS are included in the following table.

Tests of Between-Subjects Effects

Dependent Variable:HeartRate

Source	Type III Sum of Squares	df	Mean Square	F	Sig.
Corrected Model	215256.090[a]	3	71752.030	296.345	.000
Intercept	1.240E7	1	1.240E7	51206.259	.000
Group	168432.080	1	168432.080	695.647	.000
Gender	45030.005	1	45030.005	185.980	.000
Group * Gender	1794.005	1	1794.005	7.409	.007
Error	192729.830	796	242.123		
Total	1.281E7	800			
Corrected Total	407985.920	799			

a. R Squared = .528 (Adjusted R Squared = .526)

The graph appears to show that there is a main effect due to gender and group but no interaction since the lines are separate but approximately parallel. However, with large sample sizes, even small results can be significant. As seen in the table from SPSS, both gender and group have a significant effect on heart rate and there is an interaction between these factors. The means plot shows us that females have higher heart rates than males on average and that runners have lower heart rates than sedentary people. The difference between males and females is greater for the control group than for the runners.

Case Study 15.2

The side-by-side boxplots for species show species 3 looks much higher than the others (except for some overlap with species 1), and species 4 looks much lower than the others. Species 2 is also much lower than species 1 and 3.

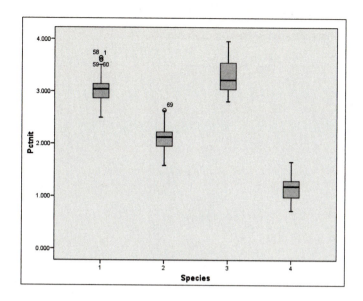

The side-by-side boxplots for water show that the distributions look fairly similar.

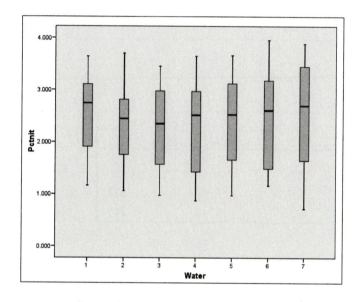

The means plot shows that species 3 is the highest for almost all water levels, followed closely by species 1. Species 2 is much lower, and species 4 is even lower. The lines are fairly parallel, but the lines for species 3 and 1 do intersect.

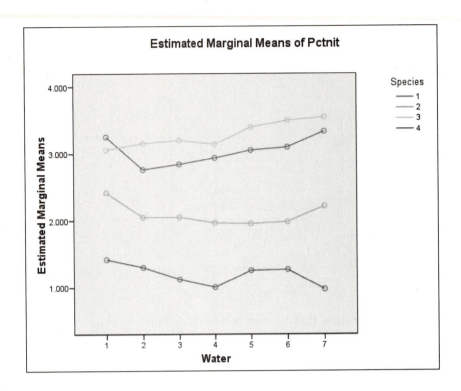

1. Species

Dependent Variable: Pctnit

Species	Mean	Std. Error	95% Confidence Interval	
			Lower Bound	Upper Bound
1	3.040	.026	2.988	3.092
2	2.093	.026	2.041	2.145
3	3.284	.026	3.232	3.337
4	1.196	.026	1.143	1.248

2. Water

Dependent Variable: Pctnit

Water	Mean	Std. Error	95% Confidence Interval	
			Lower Bound	Upper Bound
1	2.539	.035	2.470	2.608
2	2.318	.035	2.249	2.387
3	2.305	.035	2.236	2.374
4	2.264	.035	2.195	2.333
5	2.415	.035	2.346	2.484
6	2.462	.035	2.393	2.531
7	2.519	.035	2.450	2.588

The two-way ANOVA F tests show that water, species, and their interaction all are significant (with P-values close to 0). According to the Bonferroni multiple comparisons tests, all the species are significantly different from each other. (Note that none of the confidence intervals in the Bonferroni output contain 0.)

Multiple Comparisons

Dependent Variable: Pctnit

Bonferroni

(I) Species	(J) Species	Mean Difference (I-J)	Std. Error	Sig.	95% Confidence Interval Lower Bound	95% Confidence Interval Upper Bound
1	2	.94697*	.037441	.000	.84730	1.04663
	3	-.24456*	.037441	.000	-.34422	-.14489
	4	1.84422*	.037441	.000	1.74456	1.94389
2	1	-.94697*	.037441	.000	-1.04663	-.84730
	3	-1.19152*	.037441	.000	-1.29119	-1.09186
	4	.89725*	.037441	.000	.79759	.99692
3	1	.24456*	.037441	.000	.14489	.34422
	2	1.19152*	.037441	.000	1.09186	1.29119
	4	2.08878*	.037441	.000	1.98911	2.18844
4	1	-1.84422*	.037441	.000	-1.94389	-1.74456
	2	-.89725*	.037441	.000	-.99692	-.79759
	3	-2.08878*	.037441	.000	-2.18844	-1.98911

Based on observed means.

*. The mean difference is significant at the .05 level.

The Bonferroni multiple comparisons for water show that water level 1 is significantly different from water levels 2 and 3, water level 6 is significantly different from water levels 3 and 4. Water level 7 is significantly different form water levels 2, 3, and 4.

Multiple Comparisons

Dependent Variable: Pctnit
Bonferroni

(I) Water	(J) Water	Mean Difference (I-J)	Std. Error	Sig.	95% Confidence Interval	
					Lower Bound	Upper Bound
1	2	.22156*	.049530	.000	.06934	.37377
	3	.23419*	.049530	.000	.08198	.38641
	4	.27592*	.049530	.000	.12370	.42813
	5	.12397	.049530	.274	-.02824	.27618
	6	.07756	.049530	1.000	-.07466	.22977
	7	.02086	.049530	1.000	-.13135	.17307
2	1	-.22156*	.049530	.000	-.37377	-.06934
	3	.01264	.049530	1.000	-.13957	.16485
	4	.05436	.049530	1.000	-.09785	.20657
	5	-.09758	.049530	1.000	-.24980	.05463
	6	-.14400	.049530	.084	-.29621	.00821
	7	-.20069*	.049530	.001	-.35291	-.04848
3	1	-.23419*	.049530	.000	-.38641	-.08198
	2	-.01264	.049530	1.000	-.16485	.13957
	4	.04172	.049530	1.000	-.11049	.19393
	5	-.11022	.049530	.568	-.26243	.04199
	6	-.15664*	.049530	.037	-.30885	-.00443
	7	-.21333*	.049530	.001	-.36555	-.06112
4	1	-.27592*	.049530	.000	-.42813	-.12370
	2	-.05436	.049530	1.000	-.20657	.09785
	3	-.04172	.049530	1.000	-.19393	.11049
	5	-.15194	.049530	.051	-.30416	.00027
	6	-.19836*	.049530	.002	-.35057	-.04615
	7	-.25506*	.049530	.000	-.40727	-.10284
5	1	-.12397	.049530	.274	-.27618	.02824
	2	.09758	.049530	1.000	-.05463	.24980
	3	.11022	.049530	.568	-.04199	.26243
	4	.15194	.049530	.051	-.00027	.30416
	6	-.04642	.049530	1.000	-.19863	.10580
	7	-.10311	.049530	.808	-.25532	.04910
6	1	-.07756	.049530	1.000	-.22977	.07466
	2	.14400	.049530	.084	-.00821	.29621
	3	.15664*	.049530	.037	.00443	.30885
	4	.19836*	.049530	.002	.04615	.35057
	5	.04642	.049530	1.000	-.10580	.19863
	7	-.05669	.049530	1.000	-.20891	.09552
7	1	-.02086	.049530	1.000	-.17307	.13135
	2	.20069*	.049530	.001	.04848	.35291
	3	.21333*	.049530	.001	.06112	.36555
	4	.25506*	.049530	.000	.10284	.40727
	5	.10311	.049530	.808	-.04910	.25532
	6	.05669	.049530	1.000	-.09552	.20891

Based on observed means.

*. The mean difference is significant at the .05 level.

Case Study 15.3

Fbiomass

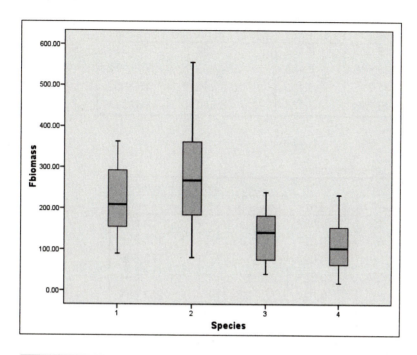

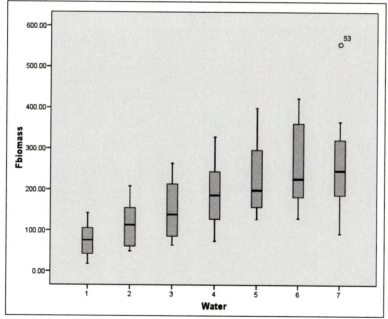

1. Species

Dependent Variable: Fbiomass

| Species | Mean | Std. Error | 95% Confidence Interval | |
			Lower Bound	Upper Bound
1	219.978	8.185	203.702	236.254
2	268.896	8.185	252.620	285.172
3	134.716	8.185	118.441	150.992
4	110.256	8.185	93.980	126.531

2. Water

Dependent Variable: Fbiomass

| Water | Mean | Std. Error | 95% Confidence Interval | |
			Lower Bound	Upper Bound
1	79.055	10.827	57.524	100.586
2	114.768	10.827	93.237	136.298
3	152.136	10.827	130.605	173.666
4	186.163	10.827	164.632	207.694
5	231.150	10.827	209.619	252.681
6	259.852	10.827	238.321	281.383
7	261.108	10.827	239.577	282.638

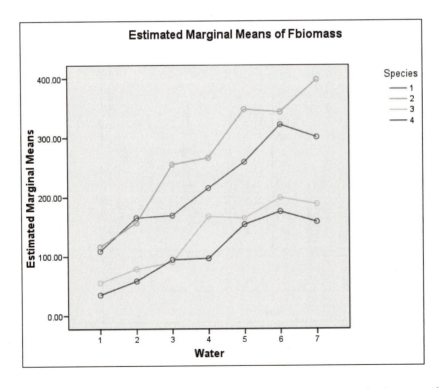

The means plot above shows that there is a fairly steady trend: As water increases, the Fbiomass also increases. Species 2 is the highest, then species 1, then species 3, and then species 4. There is little overlap of the lines. The two-way ANOVA F tests indicate that the main effect for species, the main effect for water, and their interaction are all significant. The Bonferroni multiple comparisons tests for species show that only species 3 and 4 are *not* significantly different—all the others are significantly different from each other.

Tests of Between-Subjects Effects

Dependent Variable: Fbiomass

Source	Type III Sum of Squares	df	Mean Square	F	Sig.
Corrected Model	1010577.733[a]	27	37428.805	19.956	.000
Intercept	3769710.396	1	3769710.396	2009.858	.000
Species	458295.023	3	152765.008	81.448	.000
Water	491948.413	6	81991.402	43.715	.000
Species * Water	60334.297	18	3351.905	1.787	.040
Error	157551.234	84	1875.610		
Total	4937839.363	112			
Corrected Total	1168128.967	111			

a. R Squared = .865 (Adjusted R Squared = .822)

Multiple Comparisons

Dependent Variable: Fbiomass

Bonferroni

(I) Species	(J) Species	Mean Difference (I-J)	Std. Error	Sig.	95% Confidence Interval	
					Lower Bound	Upper Bound
1	2	-48.9182*	11.57463	.000	-80.1949	-17.6415
	3	85.2614*	11.57463	.000	53.9847	116.5381
	4	109.7221*	11.57463	.000	78.4455	140.9988
2	1	48.9182*	11.57463	.000	17.6415	80.1949
	3	134.1796*	11.57463	.000	102.9030	165.4563
	4	158.6404*	11.57463	.000	127.3637	189.9170
3	1	-85.2614*	11.57463	.000	-116.5381	-53.9847
	2	-134.1796*	11.57463	.000	-165.4563	-102.9030
	4	24.4607	11.57463	.225	-6.8160	55.7374
4	1	-109.7221*	11.57463	.000	-140.9988	-78.4455
	2	-158.6404*	11.57463	.000	-189.9170	-127.3637
	3	-24.4607	11.57463	.225	-55.7374	6.8160

Based on observed means.

*. The mean difference is significant at the .05 level.

The Bonferroni output for water is shown below. Water level 2 is not significantly different from water levels 1 or 3. Water level 4 is not significantly different from water levels 3 or 5. Water levels 5, 6, and 7 are not significantly different from each other. All other water levels are significantly different from each other.

Multiple Comparisons

Dependent Variable: Fbiomass

Bonferroni

(I) Water	(J) Water	Mean Difference (I-J)	Std. Error	Sig.	95% Confidence Interval Lower Bound	95% Confidence Interval Upper Bound
1	2	-35.7125	15.31180	.464	-83.6876	12.2626
	3	-73.0806*	15.31180	.000	-121.0558	-25.1055
	4	-107.1081*	15.31180	.000	-155.0833	-59.1330
	5	-152.0950*	15.31180	.000	-200.0701	-104.1199
	6	-180.7969*	15.31180	.000	-228.7720	-132.8217
	7	-182.0525*	15.31180	.000	-230.0276	-134.0774
2	1	35.7125	15.31180	.464	-12.2626	83.6876
	3	-37.3681	15.31180	.352	-85.3433	10.6070
	4	-71.3956*	15.31180	.000	-119.3708	-23.4205
	5	-116.3825*	15.31180	.000	-164.3576	-68.4074
	6	-145.0844*	15.31180	.000	-193.0595	-97.1092
	7	-146.3400*	15.31180	.000	-194.3151	-98.3649
3	1	73.0806*	15.31180	.000	25.1055	121.0558
	2	37.3681	15.31180	.352	-10.6070	85.3433
	4	-34.0275	15.31180	.608	-82.0026	13.9476
	5	-79.0144*	15.31180	.000	-126.9895	-31.0392
	6	-107.7163*	15.31180	.000	-155.6914	-59.7411
	7	-108.9719*	15.31180	.000	-156.9470	-60.9967
4	1	107.1081*	15.31180	.000	59.1330	155.0833
	2	71.3956*	15.31180	.000	23.4205	119.3708
	3	34.0275	15.31180	.608	-13.9476	82.0026
	5	-44.9869	15.31180	.090	-92.9620	2.9883
	6	-73.6888*	15.31180	.000	-121.6639	-25.7136
	7	-74.9444*	15.31180	.000	-122.9195	-26.9692
5	1	152.0950*	15.31180	.000	104.1199	200.0701
	2	116.3825*	15.31180	.000	68.4074	164.3576
	3	79.0144*	15.31180	.000	31.0392	126.9895
	4	44.9869	15.31180	.090	-2.9883	92.9620
	6	-28.7019	15.31180	1.000	-76.6770	19.2733
	7	-29.9575	15.31180	1.000	-77.9326	18.0176
6	1	180.7969*	15.31180	.000	132.8217	228.7720
	2	145.0844*	15.31180	.000	97.1092	193.0595
	3	107.7163*	15.31180	.000	59.7411	155.6914
	4	73.6888*	15.31180	.000	25.7136	121.6639
	5	28.7019	15.31180	1.000	-19.2733	76.6770
	7	-1.2556	15.31180	1.000	-49.2308	46.7195
7	1	182.0525*	15.31180	.000	134.0774	230.0276
	2	146.3400*	15.31180	.000	98.3649	194.3151
	3	108.9719*	15.31180	.000	60.9967	156.9470
	4	74.9444*	15.31180	.000	26.9692	122.9195
	5	29.9575	15.31180	1.000	-18.0176	77.9326
	6	1.2556	15.31180	1.000	-46.7195	49.2308

Based on observed means.

*. The mean difference is significant at the .05 level.

Dbiomass

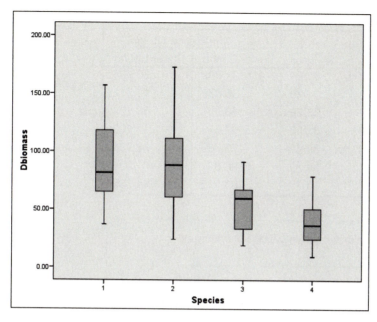

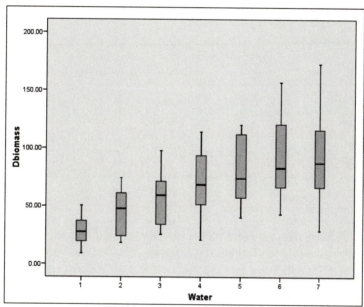

1. Species

Dependent Variable: Dbiomass

Species	Mean	Std. Error	95% Confidence Interval	
			Lower Bound	Upper Bound
1	88.862	2.743	83.407	94.317
2	85.168	2.743	79.713	90.623
3	52.204	2.743	46.749	57.659
4	39.085	2.743	33.630	44.540

2. Water

Dependent Variable: Dbiomass

Water	Mean	Std. Error	95% Confidence Interval	
			Lower Bound	Upper Bound
1	29.209	3.629	21.992	36.425
2	44.096	3.629	36.879	51.312
3	55.973	3.629	48.757	63.189
4	69.892	3.629	62.676	77.108
5	80.985	3.629	73.769	88.201
6	93.819	3.629	86.602	101.035
7	90.336	3.629	83.119	97.552

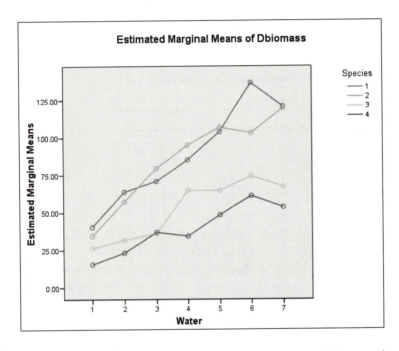

The means plot above shows that, as water increases, Dbiomass increases. Species 1 and 2 have plenty of overlap, then there is a gap, then species 3, and at the bottom, species 4. The two-way ANOVA *F* tests indicate that the main effects for species and water and their interaction are all significant. The Bonferroni multiple comparisons test shows that only species 1 and 2 are not significantly different from each other; all the other combinations of species are significantly different from each other.

Tests of Between-Subjects Effects

Dependent Variable: Dbiomass

Source	Type III Sum of Squares	df	Mean Square	F	Sig.
Corrected Model	115566.245[a]	27	4280.231	20.315	.000
Intercept	492760.264	1	492760.264	2338.736	.000
Species	50523.790	3	16841.263	79.932	.000
Water	56623.630	6	9437.272	44.791	.000
Species * Water	8418.825	18	467.712	2.220	.008
Error	17698.388	84	210.695		
Total	626024.896	112			
Corrected Total	133264.632	111			

a. R Squared = .867 (Adjusted R Squared = .825)

Multiple Comparisons

Dependent Variable: Dbiomass

Bonferroni

(I) Species	(J) Species	Mean Difference (I-J)	Std. Error	Sig.	95% Confidence Interval	
					Lower Bound	Upper Bound
1	2	3.6939	3.87939	1.000	-6.7889	14.1767
	3	36.6579*	3.87939	.000	26.1751	47.1406
	4	49.7775*	3.87939	.000	39.2947	60.2603
2	1	-3.6939	3.87939	1.000	-14.1767	6.7889
	3	32.9639*	3.87939	.000	22.4811	43.4467
	4	46.0836*	3.87939	.000	35.6008	56.5664
3	1	-36.6579*	3.87939	.000	-47.1406	-26.1751
	2	-32.9639*	3.87939	.000	-43.4467	-22.4811
	4	13.1196*	3.87939	.007	2.6369	23.6024
4	1	-49.7775*	3.87939	.000	-60.2603	-39.2947
	2	-46.0836*	3.87939	.000	-56.5664	-35.6008
	3	-13.1196*	3.87939	.007	-23.6024	-2.6369

Based on observed means.

*. The mean difference is significant at the .05 level.

The Bonferroni test for water level (given on the following page) shows that the following water levels are not significantly different: 1 and 2, 2 and 3, 3 and 4, 4 and 5, 5 and 6, 5 and 7, 6 and 7. The remaining water levels are significantly different from each other.

Multiple Comparisons

Dependent Variable: Dbiomass

Bonferroni

(I) Water	(J) Water	Mean Difference (I-J)	Std. Error	Sig.	95% Confidence Interval Lower Bound	Upper Bound
1	2	-14.8869	5.13195	.100	-30.9664	1.1926
	3	-26.7644*	5.13195	.000	-42.8439	-10.6849
	4	-40.6831*	5.13195	.000	-56.7626	-24.6036
	5	-51.7763*	5.13195	.000	-67.8557	-35.6968
	6	-64.6100*	5.13195	.000	-80.6895	-48.5305
	7	-61.1269*	5.13195	.000	-77.2064	-45.0474
2	1	14.8869	5.13195	.100	-1.1926	30.9664
	3	-11.8775	5.13195	.485	-27.9570	4.2020
	4	-25.7963*	5.13195	.000	-41.8757	-9.7168
	5	-36.8894*	5.13195	.000	-52.9689	-20.8099
	6	-49.7231*	5.13195	.000	-65.8026	-33.6436
	7	-46.2400*	5.13195	.000	-62.3195	-30.1605
3	1	26.7644*	5.13195	.000	10.6849	42.8439
	2	11.8775	5.13195	.485	-4.2020	27.9570
	4	-13.9188	5.13195	.170	-29.9982	2.1607
	5	-25.0119*	5.13195	.000	-41.0914	-8.9324
	6	-37.8456*	5.13195	.000	-53.9251	-21.7661
	7	-34.3625*	5.13195	.000	-50.4420	-18.2830
4	1	40.6831*	5.13195	.000	24.6036	56.7626
	2	25.7963*	5.13195	.000	9.7168	41.8757
	3	13.9188	5.13195	.170	-2.1607	29.9982
	5	-11.0931	5.13195	.703	-27.1726	4.9864
	6	-23.9269*	5.13195	.000	-40.0064	-7.8474
	7	-20.4438*	5.13195	.003	-36.5232	-4.3643
5	1	51.7763*	5.13195	.000	35.6968	67.8557
	2	36.8894*	5.13195	.000	20.8099	52.9689
	3	25.0119*	5.13195	.000	8.9324	41.0914
	4	11.0931	5.13195	.703	-4.9864	27.1726
	6	-12.8338	5.13195	.301	-28.9132	3.2457
	7	-9.3506	5.13195	1.000	-25.4301	6.7289
6	1	64.6100*	5.13195	.000	48.5305	80.6895
	2	49.7231*	5.13195	.000	33.6436	65.8026
	3	37.8456*	5.13195	.000	21.7661	53.9251
	4	23.9269*	5.13195	.000	7.8474	40.0064
	5	12.8338	5.13195	.301	-3.2457	28.9132
	7	3.4831	5.13195	1.000	-12.5964	19.5626
7	1	61.1269*	5.13195	.000	45.0474	77.2064
	2	46.2400*	5.13195	.000	30.1605	62.3195
	3	34.3625*	5.13195	.000	18.2830	50.4420
	4	20.4438*	5.13195	.003	4.3643	36.5232
	5	9.3506	5.13195	1.000	-6.7289	25.4301
	6	-3.4831	5.13195	1.000	-19.5626	12.5964

Based on observed means.

*. The mean difference is significant at the .05 level.

Chapter 16: Nonparametric Tests

16.1 In the table below in the "Area" row, SA = South Asia and ME = Middle East and North Africa.

Rank	1	2	3	4	5	6	7	8	9	10
Rate	3.7	3.8	3.8	4.0	4.1	4.6	4.9	5.5	5.6	5.9
Area	SA	ME	ME	ME	ME	ME	ME	ME	ME	ME

Rank	11	12	13	14	15	16	17	18	19
Rate	6.3	6.5	6.6	6.7	7.7	8.4	8.5	8.1	12.8
Area	SA	SA	ME	SA	SA	ME	SA	SA	SA

The Middle East and North Africa rates have lower ranks than the South Asia rates.

16.3 Using the Normal approximation with the continuity correction for a one-tailed test, the test statistic is $Z = 2.19$, and the P-value = 0.0143. These results are close to the results from SPSS, and the same conclusion applies.

16.5 The table below shows the mpg, average rank, and number of occurrences of each rank.

MPG	13	15	16	17	18	19	20	21
Rank	1.5	3.5	9.5	19.5	27	32	38.5	46
Occurences	2	2	10	10	5	5	8	7

MPG	22	23	24	25	26	27	28	29	30
Rank	50.5	52.5	56	60	62	65	68	69	70
Occurences	2	2	5	3	1	5	1	1	1

16.7 Yes, there is strong evidence that healthy firms have a higher ratio of assets to liabilities on the average. $W_{healthy} = 4299$, and $W_{failed} = 852$, $Z = -6.017$, and the P-value from SPSS for the one sided test is very close to 0.

16.9 H_0: There is no difference in the ego strength of the high-fitness and low-fitness groups; H_a: Ego strength is significantly different between the two fitness groups. $W_{Low} = 105$, $W_{High} = 301$, $Z = -4.503$, and the one-sided P-value from SPSS is approximately 0. There is evidence that the high-fitness group has higher ego strength than the low-fitness group.

16.11 **a)**

Price	6.8250	7.0275	7.0825	7.3000	7.3025	7.3125	7.3325	7.3600	7.5550	7.5575
Rank	1	2	3	4	5	6	7	8	9	10

b) $W_{September} = 40$. $\mu_W = 27.5$, $\sigma_W = 4.787$. **c)** P-value = 0.012.

16.13 a)

```
      8 | 09 |
        | 10 |
 0  0 | 11 | 8
      4 | 12 | 0  6   6  9
        | 13 |
      0 | 14 |
```

b) We do not have strong evidence that breaking strengths are lower for strips buried longer ($P = 0.1467$).

16.15 a) Yes, there appears to be a difference in species counts.

```
              |  0 |
              |  0 | 4
              |  0 |
              |  1 | 0  2
 5  5  3  3  3 |  1 | 4  5  5
        9  9  8 |  1 | 7  8  8
     2  2  1  0 |  2 |
              |  2 |
              |  2 |
```

b) H_0: There is no difference in the number of tree species in unlogged forests and logged forests. H_a: Logged forests have a significantly lower number of tree species. $W = 159$ with a P-value $= 0.029$. The results show a significantly lower number of tree species in logged forests.

16.17 Yes, it appears that women are more concerned than men about food safety in restaurants. $W = 32267.5$ with a P-value $= 0.0001$.

16.19 a) $\chi^2 = 3.955$, df $= 4$, P-value $= 0.412$. There is no indication of a relationship between income and city. **b)** No, there is no strong evidence to indicate a systematically higher income level in one city over the other. ($W = 56,370$, P-value $= 0.662$ with continuity correction.)

16.21 $\mu_{W^+} = 10.5$. $\sigma_{W^+} = 4.7697$. Since the alternative hypothesis for Exercise 16.20 is two sided, the P-value $= 2 \times P(W^+ \geq 21) = 2 \times P(Z \geq 2.20) = 0.0278$. There is strong evidence that there is a systematic difference between the online book prices at Barnes and Noble and at Amazon.

16.23 H_0: There is no difference between the estimates from the two garages; H_a: Jocko's garage gives higher estimates than the "trusted" garage. $W^+ = 49.5$, $\mu_{W^+} = 27.5$, and $\sigma_{W^+} = 9.8107$. The test statistic is $Z = 2.24$, resulting in a P-value of 0.0125. There is evidence that Jocko's garage has higher estimates than the "trusted" garage.

16.25 Using SPSS, the test statistic is $Z = -0.595$, and the P-value is 0.552 for the two-sided test. There is not enough evidence to say that the distribution of battery life for MP3 players has a population median different from 6.5.

16.27 H_0: median increase$_{low}$ = 0; H_a: median increase$_{low}$ > 0. The Wilcoxon signed rank test results gives $\mu_{W+} = 5, \sigma_{W+} = 2.74$, $W^+ = 10$, P-value = $P(W^+ > 10) = P(Z > 1.642) \approx$ 0.05. This test is just significant at the 0.05 level.

16.29 a)

Score	1	1	2	2	2	3	3	3	3	3	3	6	6	6	6	6	**6**
Rank	1.5	1.5	4	4	4	8.5	8.5	8.5	8.5	8.5	8.5	14.5	14.5	14.5	14.5	14.5	**14.5**

$W^+ = 138.5$. The bold-faced column represents the one negative value. **b)** H_0: The distributions of the test scores are the same before and after a course. H_a: The scores are systematically higher after the course. The P-value = 0.002. Conclude that the scores are higher after the course.

16.31 P-value = 0.206. There is no strong evidence of a systematic difference between the perceived safety of fast food and fair food.

16.33 $W^+ = 88$ with a P-value = 0.059. The evidence is not significant at the 0.05 level.

16.35 Minitab gives an estimated median of 36.5 with a 95% confidence interval of (28.0, 44.5).

16.37 a) $H = 8.73$ with a P-value = 0.068. The results of this test are not significant at the 0.05 level. It appears that there is not a significant difference in vitamin C loss over time. **b)** Yes, the difference in P-values would lead to different conclusions.

16.39 a) The ANOVA tests the hypothesis that the four means are the same. The Kruskal-Wallis tests the hypothesis that the four distributions are the same.
b) The Minitab output containing the medians and the H statistic is below. Lemon Yellow appears most effective. The results of the test indicate a significant difference in the number of insects attracted by each color.

Kruskal-Wallis Test

```
C9              N      Median    Ave Rank        Z
Blue            6       15.00         6.7     -2.33
Green           6       34.50        14.8      0.93
LYellow         6       46.50        21.2      3.47
White           6       15.50         7.3     -2.07
Overall        24                    12.5

H = 16.95    DF = 3    P = 0.001
H = 16.98    DF = 3    P = 0.001 (adjusted for ties)
```

16.41 a) 4.6, 6.54, 9.53, 16.09. **b)** 126, 126, 136, 110. The hypothesis tested is that the medians are all equal. **c)** $H = 5.631$ with a P-value = 0.131. The conclusion is that there is not a significant difference in the decay medians between the four lengths of time.

16.43   a)

Unlogged	Logged(1)	Logged(8)
13 000	0 2	0 4
14	0	0
15 00	0 7	0
16	0 8	1 0
17	1 11	1 2
18 0	1 23	1 455
19 00	1 4555	1 7
20 0	1	1 88
21 0	1 8	
22 00		

The stemplots show many outliers for each of the distributions. The medians are 18.5, 12.5, and 15. **b)** $H = 9.44$ with a P-value $= 0.009$. The conclusion would be that there is a significant difference in medians between the three groups.

16.45   a) Verizon does not appear to give CLEC customers the same level of service as its own customers. The average time to respond to trouble calls is 1.73 hours for Verizon's customers and 3.80 hours for CLEC's customers. Also of note is that the distribution for Verizon's customers is skewed to the right with the vast majority of calls taking approximately one hour. CLEC's customers are more likely to wait approximately 5 hours, while very few Verizon customers wait this long.

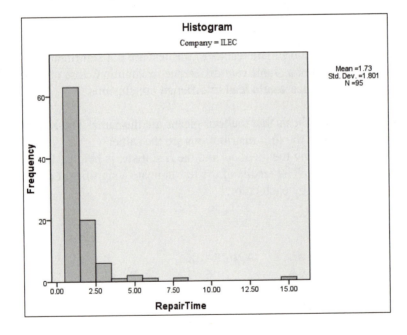

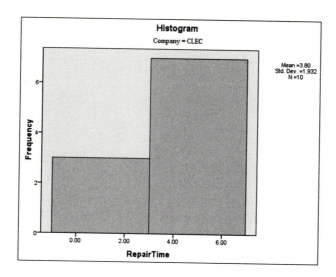

b) We might hesitate to use a *t* test due to the strong skewness of the data, only integer values being used for the responses, and the fact that the sample sizes are so different, including one very small sample size. $W_{\text{Verizon}} = 4778.50$, $W_{\text{CLEC}} = 786.50$. $Z = -3.246$, *P*-value = 0.0005. There is strong evidence that the time to respond to trouble calls is greater for CLEC's customers than for Verizon's customers.

16.47 **a)** See the graphical and numerical summaries below. The median for control is 0.219, the median for low dose is 0.216, and the median for high dose is 0.232. Control is right skewed, low dose is fairly symmetric, and high dose is slightly right skewed. The high dose distribution appears to be much higher than the other two distributions in general. There is quite a bit of overlap between the control and low dose distributions. **b)** H_0: The bone mineral densities have the same distribution in all treatment groups, H_a: The bone mineral densities are systematically higher in some treatment groups than in others. SPSS gives a test statistic of $H = 9.116$ and a *P*-value = 0.010. There is evidence that the bone mineral densities are systematically higher in some treatment groups than in others.

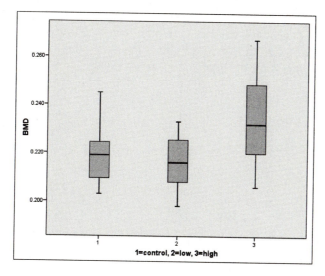

Descriptives

BMD

	N	Mean	Std. Deviation	Std. Error	95% Confidence Interval for Mean		Minimum	Maximum
					Lower Bound	Upper Bound		
1	15	.21887	.011587	.002992	.21245	.22528	.203	.245
2	15	.21593	.011511	.002972	.20956	.22231	.198	.233
3	15	.23507	.018771	.004847	.22467	.24546	.206	.267
Total	45	.22329	.016413	.002447	.21836	.22822	.198	.267

16.49   a)

Beef hot dogs

```
 8
 9
10
11 1
12
13 1259
14 1899
15 2378
16
17 56
18 146
19 00
```

Meat hot dogs

```
 8
 9
10 7
11
12
13 5689
14 067
15 3
16
17 2359
18 2
19 015
```

Poultry hot dogs

```
 8 67
 9 49
10 226
11 3
12 9
13 25
14 2346
15 2
16
17 0
18
19
```

The five-number summaries are:

Hot Dog Type	Minimum	Q_1	Median	Q_3	Maximum
Beef	111	139.5	152.5	179.8	190
Meat	107	138.5	153.0	180.5	195
Poultry	86	100.5	129.0	143.5	170

b) The distributions of the beef and meat hot dogs appear non-Normal. They are not symmetrical, have a gap in the data at 160 calories, and both have one low outlier. The poultry hot dogs have one high outlier. **c)** $H = 15.89$ with a P-value $= 0.000$. There is a systematically higher calorie content for the beef and meat hot dogs.

16.51   a)

Beef hot dogs

```
1
1
2
2 59
3 011223
3 778
4 024
4 7789
5
5 8
6 4
```

Meat hot dogs

```
1 4
1
2
2
3 3
3 67889
4 002
4 579
5 0014
5
6
```

Poultry hot dogs

```
1
1
2
2
3
3 557889
4 23
4
5 112244
5 8
6
```

Hot Dog Type	Minimum	Q_1	Median	Q_3	Maximum
Beef	253	319.8	380.5	478.5	645
Meat	144	379.0	405.0	501.0	545
Poultry	357	379.0	430.0	535.0	588

b) The distributions all have outliers and the poultry sodium distribution appears to have two distinct groups of data. **c)** $H = 4.71$ with a *P*-value = 0.095. The evidence of systematically higher sodium content for some types of hot dogs is not strong.

16.53 a) Let W_1 be the sum of the ranks from n_1. Let W_2 be the sum of the ranks from n_2.

$$W_1 + W_2 = \sum_{i=1}^{N} i. \quad \mu_{W_1} + \mu_{W_2} = \mu_{W_1+W_2} = \mu_{\sum_{i=1}^{N} i} = \sum_{i=1}^{N} i.$$

$$\sum_{i=1}^{N} i = \frac{n_1(N+1)}{2} + \frac{n_2(N+1)}{2} = \frac{N(N+1)}{2}.$$

b) For $N = 27$, $(27 \times 28)/2 = 378$. **c)** $(62 \times 63)/2 = 1953$. $308 + 350 + 745 + 550 = 1953$.

Chapter 17: Logistic Regression

17.1 For exclusive territory, the proportion = 0.7606, and the odds = 3.177. For non-exclusive territory, the proportion = 0.536, and the odds = 1.155.

17.3 1.156, 0.144.

17.5 $b_0 = 0.143$, $b_1 = 1.013$. log(ODDS) = 0.143 + 1.013x. The odds ratio of exclusive territories to non-exclusive territories is 2.754.

17.7 $$\frac{ODDS_{x+1}}{ODDS_x} = \frac{e^{-11.0391}e^{3.1709(x+1)}}{e^{-11.0391}e^{3.1709x}} = \frac{e^{3.1709x}e^{3.1709}}{e^{3.1709x}} = e^{3.1709} = 23.829.$$

17.9 log (ODDS) = 0.1431 + 1.0127x, where $x = 1$ if the franchise has an exclusive territory and $x = 0$ otherwise. The odds ratio of exclusive territories to nonexclusive territories is 2.75 with a 95% confidence interval of (1.19, 6.36).

17.11 (2.3485, 3.8691).

17.13 Verify.

17.15 **a)** The odds of the event is increased by $e^2 = 7.39$ when the explanatory variable increases by 1. **b)** The intercept is equal to the log odds of an event when $x = 0$. **c)** The odds of an event is the probability of the event divided by one minus the probability of the event (or the probability of success divided by the probability of failure).

17.17 **a)** 0.1765. **b)** 0.5584. **c)** 0.7967.

17.19 Senior adults and those who speak English as a second language have odds of about 3 to 4 of leaving a high tip. The odds that a French-speaking Canadian will leave a high tip are about 4 to 5. The odds that someone ordering alcoholic drinks will leave a high tip are about 9 to 8. So, a customer who is a senior adult, who speaks English as a second language or who is French-speaking Canadian will be significantly less likely to leave a high tip than an average customer, but those served alcohol will be significantly more likely to leave a high tip.

17.21 **a)** The confidence intervals for the odds ratios are related to testing the null hypothesis that the odds ratio = 1 versus the two-sided alternative. If the confidence interval contains 1, the null hypothesis should not be rejected. **b)** For reader age, model sex, and women's magazines, 1 is outside the confidence interval, and therefore the null hypothesis should be rejected. Only the confidence interval for men's magazines contains 1, and therefore we would not reject the null hypothesis for this odds ratio. **c)** The odds ratio confidence intervals show that men's magazines has no effect on the probability that an ad will be not sexual. All the other variables do have an effect on the probability that an ad will not be sexual. The confidence interval for the odds ratio and the Wald test yield the same results. It is generally easier to explain the results using odds ratios than the coefficients from the model though.

17.23 **a)** A person is more likely to use the Internet if they are in an urban location, have at least some postsecondary education, or speak English. Males are less likely to use the Internet than females. As age increases, a person is less likely to use the Internet, but as income increases, a person is more likely to use the Internet. **b)** Odds ratios are 0.94 for age, 1.01 for income, 1.44 for location, 0.80 for sex, 2.94 for education, 1.33 for language, and 1.05 for children. **c)** 10.05.

17.25 **a)** 0.80, 4 to 1. **b)** 0.69, 2.23. **c)** 1.84.

17.27 **a)** (−0.047, 1.265). **b)** (0.95, 3.54). **c)** There does not seem to be a difference in the proportions of high tech companies and non-high tech companies that offer stock options.

17.29 **a)** 0.21. **b)** 0.27. **c)** 0.79. **d)** 3.67. **e)** They are reciprocals of each other.

17.31 **a)** proportion = 0.0165, odds = 0.0168 or approximately 2 to 100. **b)** proportion = 0.0078, odds = 0.0079. **c)** 2.12. Men with high-blood pressure are 2.12 times more likely to die of cardiovascular disease than men with low-blood pressure.

17.33 **a)** 2.12, (1.28, 3.51). **b)** There is evidence that the odds are higher that a man with high blood pressure will die from cardiovascular disease than that a man with low blood pressure will die from this disease.

17.35 **a)** $\log(\text{ODDS}) = -3.502 + 1.037x$. **b)** For a binomial distribution, we would assume that each employee is an independent trial and each employee has the same chance of being terminated. This may not be realistic because the employee's performance is likely to play a role in their termination. **c)** 2.82, (1.64, 4.84). An employee that is over 40 years old is 2.82 times more likely to be terminated than an employee that is 40 or younger. The confidence interval on the odds estimate is shows that the odds are significantly greater than 1. **d)** You can also incorporate performance evaluations as an explanatory variable by creating a multiple logistic regression model.

17.37 $\log(\text{ODDS}) = -0.0282 + 0.6393x$. $\chi^2 = 48.30$ with a P-value = 0.000. The odds ratio estimate is 1.90. There are significantly greater odds for a college graduate to use the Internet to make travel arrangements than a non-college graduate.

17.39 **a)** The model is $\log\left(\dfrac{p}{1-p}\right) = -2.465 + 1.038x_{over40}$ (with $x_{over40} = 0$ if not over 40, $x_{over40} = 1$ if over 40). The coefficient for x_{over40} is significantly different from 0. If a person is not over 40, $\log\left(\dfrac{p}{1-p}\right)$ is −2.465 with an estimated odds ratio of 0.030. If a person is over 40, $\log\left(\dfrac{p}{1-p}\right)$ is −2.465 with an estimated odds ratio of 0.085.

b) Answers will vary, but other variables might include performance ratings, years of service to the company, and education level.

17.41 Verify.

17.43 **a)**

Division	x_{II}	x_{III}	$\log\left(\dfrac{p}{1-p}\right)$	Estimated $\dfrac{p}{1-p}$
I	0	0	−1.572	0.208
II	1	0	−1.325	0.266
III	0	1	−1.130	0.323

b) See the scatterplot below. The trend is extremely linear. **c)** If simple linear regression is used, the fit is extremely good with an $R^2 = 99.5\%$. The equation of the line is
$$\hat{y} = -1.784 + 0.221 x_{division}.$$

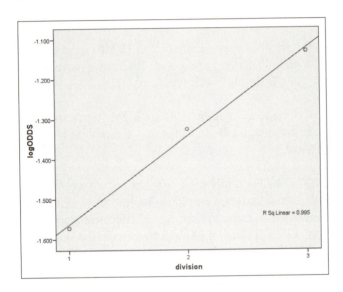

Model Summary

Model	R	R Square	Adjusted R Square	Std. Error of the Estimate
1	.998[a]	.995	.991	.021229

a. Predictors: (Constant), division

Coefficients[a]

Model		Unstandardized Coefficients		Standardized Coefficients	t	Sig.
		B	Std. Error	Beta		
1	(Constant)	-1.784	.032		-55.025	.012
	division	.221	.015	.998	14.722	.043

a. Dependent Variable: logODDS

Case Study 17.1

Lactic: The model is $\log\left(\dfrac{p}{1-p}\right) = -10.780 + 6.332x_{lactic}$. The odds ratio for b_{lactic} is 562.278. If we increase the lactic acid content by one unit, we increase the odds that the cheese will be acceptable by 562.278. The coefficient for lactic acid is statistically significant.

Variables in the Equation

		B	S.E.	Wald	df	Sig.	Exp(B)	95.0% C.I.for EXP(B) Lower	Upper
Step 1[a]	Lactic	6.332	2.453	6.662	1	.010	562.240	4.590	68874.866
	Constant	-10.780	3.975	7.353	1	.007	.000		

a. Variable(s) entered on step 1: Lactic.

H2S: The model is $\log\left(\dfrac{p}{1-p}\right) = -7.279 + 0.940x_{H2S}$. The odds ratio for b_{H2S} is 2.600. If we increase the H2S content by one unit, we increase the odds that the cheese will be acceptable by 2.600. The coefficient for H2S is statistically significant.

Variables in the Equation

		B	S.E.	Wald	df	Sig.	Exp(B)	95.0% C.I.for EXP(B) Lower	Upper
Step 1[a]	H2S	.940	.344	7.451	1	.006	2.560	1.303	5.027
	Constant	-7.279	2.522	8.332	1	.004	.001		

a. Variable(s) entered on step 1: H2S.

Lactic and H2S: The model is $\log\left(\dfrac{p}{1-p}\right) = -11.718 + 3.777x_{lactic} + 0.735x_{H2S}$. The chi-square test for logistic regression has a test statistic of 16.192 with 2 degrees of freedom and a P-value close to zero, so at least one of these explanatory variables can be used to predict the odds that the cheese is acceptable. The tests for the individual coefficients at the 5% level shows that neither of these explanatory variables adds significant predictive ability once the other one is already in the model.

Omnibus Tests of Model Coefficients

		Chi-square	df	Sig.
Step 1	Step	16.192	2	.000
	Block	16.192	2	.000
	Model	16.192	2	.000

Variables in the Equation

		B	S.E.	Wald	df	Sig.	Exp(B)	95.0% C.I.for EXP(B) Lower	Upper
Step 1[a]	Lactic	3.777	2.596	2.116	1	.146	43.679	.269	7084.337
	H2S	.735	.387	3.612	1	.057	2.085	.977	4.447
	Constant	-11.718	4.437	6.973	1	.008	.000		

a. Variable(s) entered on step 1: Lactic, H2S.

Lactic and Acetic: The model is $\log\left(\dfrac{p}{1-p}\right) = -16.558 + 5.257x_{lactic} + 1.309x_{acetic}$.

The chi-square test for logistic regression has a test statistic of 12.877 with 2 degrees of freedom and a *P*-value of 0.002, so at least one of these explanatory variables can be used to predict the odds that the cheese is acceptable. The tests for the individual coefficients at the 5% level shows that lactic acid adds significant predictive ability when acetic acid is already in the model, but the same cannot be said of acetic acid when lactic acid is already in the model.

Omnibus Tests of Model Coefficients

		Chi-square	df	Sig.
Step 1	Step	12.877	2	.002
	Block	12.877	2	.002
	Model	12.877	2	.002

Variables in the Equation

		B	S.E.	Wald	df	Sig.	Exp(B)	95.0% C.I.for EXP(B) Lower	Upper
Step 1[a]	Lactic	5.257	2.500	4.421	1	.036	191.939	1.428	25795.224
	Acetic	1.309	1.288	1.033	1	.309	3.702	.297	46.181
	Constant	-16.558	7.422	4.977	1	.026	.000		

a. Variable(s) entered on step 1: Lactic, Acetic.

H2S and Acetic: The model is $\log\left(\dfrac{p}{1-p}\right) = -12.847 + 1.096x_{acetic} + 0.830x_{H2S}$. The chi-square test for logistic regression has a test statistic of 14.226 with 2 degrees of freedom and a *P*-value of 0.001, so at least one of these explanatory variables can be used to predict the odds that the cheese is acceptable. The tests for the individual coefficients at the 5% level shows that H2S adds significant predictive ability when acetic acid is already in the model, but the same cannot be said of acetic acid when H2S is already in the model.

Omnibus Tests of Model Coefficients

		Chi-square	df	Sig.
Step 1	Step	14.226	2	.001
	Block	14.226	2	.001
	Model	14.226	2	.001

Variables in the Equation

		B	S.E.	Wald	df	Sig.	Exp(B)	95.0% C.I.for EXP(B) Lower	Upper
Step 1ᵃ	Acetic	1.096	1.382	.629	1	.428	2.993	.199	44.926
	H2S	.830	.367	5.109	1	.024	2.294	1.117	4.712
	Constant	-12.847	7.868	2.666	1	.102	.000		

a. Variable(s) entered on step 1: Acetic, H2S.

Case Study 17.2

The chi-square test for multiple logistic regression has a test statistic of 37.197 with 5 degrees of freedom and a *P*-value close to 0, so we conclude that one or more of these explanatory variables can be used to predict the odds that the GPA will be at least 3.0. The model is

$$\log\left(\frac{p}{1-p}\right) = -7.373 + 0.343x_{HSM} + 0.225x_{HSS} + 0.019x_{HSE} + 0.001x_{SATM} + 0.003x_{SATV}.$$

The tests for the individual coefficients show that, at the 5% level, only HSM adds significant predictive ability once the other variables are already in the model. (At the 10% significance level, HSS would also add significant predictive ability.)

In Exercises 11.111 to 11.119, HSM had the only coefficient significantly different from 0. This agrees with the results from the multiple logistic regression.

Omnibus Tests of Model Coefficients

		Chi-square	df	Sig.
Step 1	Step	37.197	5	.000
	Block	37.197	5	.000
	Model	37.197	5	.000

Variables in the Equation

		B	S.E.	Wald	df	Sig.	Exp(B)	95.0% C.I.for EXP(B) Lower	Upper
Step 1ᵃ	HSM	.343	.142	5.834	1	.016	1.409	1.067	1.861
	HSS	.225	.129	3.055	1	.080	1.252	.973	1.611
	HSE	.019	.129	.022	1	.883	1.019	.792	1.312
	SATM	.001	.002	.106	1	.745	1.001	.996	1.005
	SATV	.003	.002	2.280	1	.131	1.003	.999	1.007
	Constant	-7.373	1.477	24.926	1	.000	.001		

a. Variable(s) entered on step 1: HSM, HSS, HSE, SATM, SATV.

Case Study 17.3

For the combined data set (not separating out patients by condition and using Hospital A = 0, Hospital B = 1), the model is $\log\left(\frac{p}{1-p}\right) = -3.476 - 0.416x_{hospital}$, and the coefficient for hospital is not significantly different from 0. The 95% confidence interval for the odds ratio for hospital is (0.379, 1.149).

Variables in the Equation

		B	S.E.	Wald	df	Sig.	Exp(B)	95.0% C.I.for EXP(B) Lower	Upper
Step 1[a]	hospital	-.416	.283	2.157	1	.142	.660	.379	1.149
	Constant	-3.476	.128	738.408	1	.000	.031		

a. Variable(s) entered on step 1: hospital.

When the patient condition is also taken into account with hospital (using Poor Condition = 0, Good Condition = 1), the chi-square test for multiple logistic regression has a test statistic of 21.019 with 2 degrees of freedom and a *P*-value close to 0, so there is evidence that at least one of the explanatory variables can be used to predict the odds that a patient will die. The coefficient for hospital is not significantly different from 0, which means only condition adds significant predictive ability once hospital is already in the model.

The 95% confidence interval for the odds ratio for hospital is (0.624, 2.086) and for condition is (0.150, 0.530).

When only hospital is used, the coefficient for hospital is negative. When hospital and condition are used, the coefficient for hospital is positive. For the hospital-only results, switching from hospital A to hospital B decreases a patient's chance of survival. For the hospital and condition are used, switching from hospital A to hospital B increases a patient's chance of survival.

Omnibus Tests of Model Coefficients

		Chi-square	df	Sig.
Step 1	Step	21.019	2	.000
	Block	21.019	2	.000
	Model	21.019	2	.000

Variables in the Equation

		B	S.E.	Wald	df	Sig.	Exp(B)	95.0% C.I.for EXP(B) Lower	Upper
Step 1[a]	hospital	.132	.308	.184	1	.668	1.141	.624	2.086
	condition	-1.266	.322	15.480	1	.000	.282	.150	.530
	Constant	-3.241	.133	596.304	1	.000	.039		

a. Variable(s) entered on step 1: hospital, condition.

Case Study 17.4

		Zip Code	47904	47906
Price		Mean	94,900	194,158
		St. Dev.	31,030.91	96,794.11
		Median	87,450	167,500
		Minimum	52,000	63,900
		Maximum	199,500	625,000
Square Feet		Mean	1,308.48	2,076.32
		St. Dev.	428.903	671.730
		Median	1,290	2,000
		Minimum	698	936
		Maximum	2,296	4,840

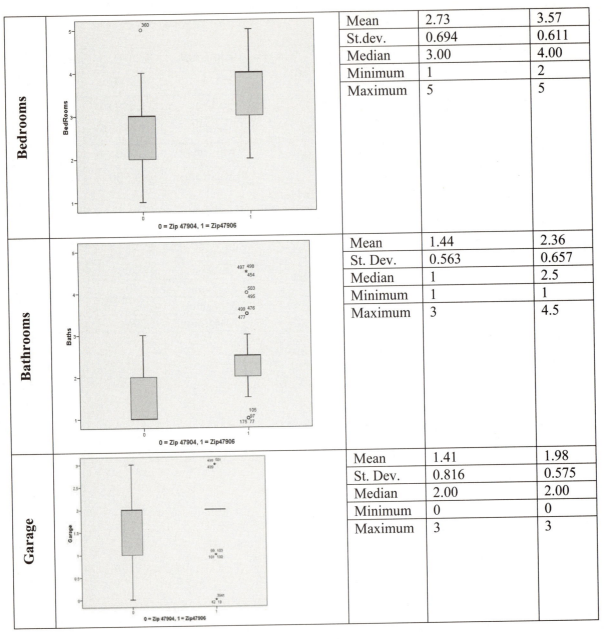

	Mean	2.73	3.57
Bedrooms	St.dev.	0.694	0.611
	Median	3.00	4.00
	Minimum	1	2
	Maximum	5	5

	Mean	1.44	2.36
Bathrooms	St. Dev.	0.563	0.657
	Median	1	2.5
	Minimum	1	1
	Maximum	3	4.5

	Mean	1.41	1.98
Garage	St. Dev.	0.816	0.575
	Median	2.00	2.00
	Minimum	0	0
	Maximum	3	3

Using all five explanatory variables, the chi-square test for the multiple logistic regression has a test statistic of 106.170, 5 degrees of freedom, and a *P*-value close to 0, so at least one of the explanatory variables can be used to predict the odds that a house is in the 47906 zip code instead of 47904.

Omnibus Tests of Model Coefficients

		Chi-square	df	Sig.
Step 1	Step	106.170	5	.000
	Block	106.170	5	.000
	Model	106.170	5	.000

The model is:

$$\log\left(\frac{p}{1-p}\right) = -11.750 + 6.56 \times 10^{-5}\, x_{price} - 0.003 x_{sqft} + 2.706 x_{bed} + 0.324 x_{bath} + 0.291 x_{garage}$$

When looking at the individual coefficients, price, square feet, and bedrooms have coefficients significantly different from 0, but bathrooms and garages do not. (The coefficient for price is not actually 0 but is so small that SPSS doesn't show enough decimal places in this output. The actual coefficient is 6.56×10^{-5}.)

Variables in the Equation

		B	S.E.	Wald	df	Sig.	Exp(B)	95.0% C.I.for EXP(B) Lower	Upper
Step 1[a]	Price	.000	.000	16.793	1	.000	1.000	1.000	1.000
	SqFt	-.003	.001	6.818	1	.009	.997	.995	.999
	BedRooms	2.706	.784	11.916	1	.001	14.976	3.221	69.628
	Baths	.324	.641	.256	1	.613	1.383	.394	4.854
	Garage	.291	.457	.406	1	.524	1.338	.546	3.275
	Constant	-11.750	2.636	19.876	1	.000	.000		

a. Variable(s) entered on step 1: Price, SqFt, BedRooms, Baths, Garage.